Asian Perspectives on Water Policy

Asian countries are not homogenous. They are in different stages of social and economic development, with cultural conditions and institutional and legal frameworks varying from one country to another. Therefore, how water can be successfully managed differs from one country to another. This book provides authoritative analyses of how water is being managed in different Asian countries, ranging from the world's most populous countries like China and India to a city state like Singapore and an island country like Fiji. It also analyses in depth several wide ranging issues like terrorism, human rights, water-energy nexus, and roles of media, along with comprehensive discussions of legal, institutional and regulatory frameworks in an Asian water management context. The overall focus is how water can be managed efficiently, cost-effectively and equitably in various Asian countries.

This book was based on a special issues of *International Journal of Water Resources Development*.

Cecilia Tortajada is the Scientific Director of the International Centre of Water and Environment, Zaragoza, Spain, and Visiting Professor at the Lee Kuan Yew School of Public Policy, National University of Singapore.

Asit K. Biswas is the Founder and Chief Executive of the Third World Centre for Water Management, an innovative think tank in Mexico. He is a Distinguished Visiting Professor at the Lee Kuan Yew School of Public Policy, National University of Singapore, and also at the Indian Institute of Technology, Bhubaneswar.

Asian Perspectives on Water Policy

Edited by
Cecilia Tortajada and Asit K. Biswas

Routledge
Taylor & Francis Group

LONDON AND NEW YORK

Lee Kuan Yew
School of Public Policy

National University of Singapore

Contents

Preface

CECILIA TORTAJADA & ASIT K. BISWAS

Asia is a vast continent, containing over 60 percent of the global population in 2010. The Asian countries are very heterogeneous, ranging from highly developed countries, like Japan and Singapore, to some of the least developed countries of the world. Social, cultural and political conditions are often radically different, as are the effectivenesses of their institutions, legal and regulatory frameworks and governance practices. The countries also have very different physical and climatic conditions, ranging from some of the rainiest places in the world to absolute deserts. Under such varying social, cultural, economic, political and physical conditions, differing institutional capacities and management practices, it is not possible to either define or to consider a common Asian water landscape.

Not only the type, magnitude and extent of the water issues vary from one Asian country to another but also it is often difficult to define a common panorama of water problems and their solutions within large countries like China, India, Indonesia or Malaysia. Whereas some Asian cities like Beijing and Shanghai can now compete or even exceed important cities of the western world in terms of availability and conditions of infrastructure, the hinterlands often present a completely different landscape. Thus, there are no common water problems of Asia as a whole, let alone common solutions to the problems that would be equally applicable over the continent. As the first Prime Minister of the independent India, Jawaharlal Nehru, perceptibly noted, if one said anything about India, it may be correct. However, if exactly the reverse is said, it is also likely to be correct but at a different location and context.

Projections of future population and consumption trends indicate that satisfying the demands for water of appropriate quantity and quality for various uses will be an important social and economic consideration for all the Asian countries in the future. Drivers such as population, urbanization, industrialization and economic development, and corresponding changes in the demands for food, energy and environmental protection are just a few of the issues that will radically affect existing and future water policies, including planning and management processes as well as water allocation patterns.

Absence of good water policies, and their implementation, has contributed to continued mismanagement of this resource in nearly all Asian countries. Extensive policy failures and institutional constraints have received limited corrective actions from the concerned governments over the recent decades. Rapid urbanization, structure of population growth, uneven economic developments, migration between and within countries, industrialization and increasing societal expectations have imposed complex and conflicting demands not only on the national water resources but also on the human capacities and institutional abilities to respond to such needs efficiently and in a timely manner. An important net result of these developments has been misuse and over-exploitation of water in nearly all the Asian countries.

There is now an urgent need to formulate and implement forward-looking, business-unusual policies that can reform and strengthen public institutions that are responsible for planning and managing water resources, increase public and private sector investments in water-related areas, manage and improve urban and rural environment, encourage use of available and appropriate technologies, seriously consider South-South knowledge, technology and experience transfer, and develop a new generation of capable policy-makers and managers with good governance and communication skills.

There is no question that as the twenty-first century progresses, the water profession will confront a problem the magnitude and complexity of which no earlier generation had to face. The profession at present has two stark choices: carry on as before with a business-as-usual and solution-in-search-of-a-problem mindset and endow future generations with a legacy of poor water policies and governance practices and plethora of partially resolved water-related problems, or continue in earnest an accelerated effort to identify, understand and then efficiently manage the current and the future water-related problems of the Asian countries.

Within this overall framework, several specialists from different sectors and disciplines associated with the Lee Kuan Yew School of Public Policy, Singapore, have reviewed, assessed and analyzed different water-related problems from various Asian countries. All but three of the papers in this book were initially published as a Special Issue of the *International Journal of Water Resources Development* in March 2010. The two papers by the Editors were published in *Hydrological Sciences Bulletin* in June 2011. Together, these papers show how different water issues and their potential solutions are from one Asian country to another, and also how these problems and solutions appear from the perspectives of different disciplines. We are confident that the readers of this book will find the papers thought provoking and stimulating.

Cooperation or Conflict in Transboundary Water Management: Case Study of South Asia

ASIT K. BISWAS

Third World Centre for Water Management, Atizapan, Mexico, and Lee Kuan Yew School of Public Policy, Singapore

ABSTRACT   *The transboundary Himalayan rivers flowing through Bhutan, Nepal, India and Bangladesh provide a golden opportunity to improve the standard of living of the largest concentration of the poor people in the world. Bhutan and India have shown that, given goodwill and trust between the countries concerned, water can be successfully used as an engine for economic growth. This can bring substantial benefits to the people of both the countries. In contrast, lack of trust between Nepal, India and Bangladesh has compounded the deprivation of the region through underdevelopment. This paper analyses two very contrasting results of managing transboundary rivers in South Asia, a most successful one in Bhutan and India, and a missed opportunity in Nepal, India and Bangladesh.*

Introduction

The management of transboundary rivers has become an important social and political issue in recent years, for a variety of reasons, some valid and others due to linear but erroneous thinking. There are several valid reasons. First, there are many major transboundary rivers and lakes where there are no treaties for water allocation between all the co-basin countries that could provide a guiding framework for water planning and management. Second, even though the Convention on the Law of Non-Navigational Uses of International Watercourses was overwhelmingly approved on 21 May 1997 by the United Nations (UN) General Assembly, with only three dissenting votes but 33 abstentions (Biswas, 2008a), it has not yet entered into force even 14 years after the initial approval. In recent years (from 2007), there appears to be a slightly increased momentum for its ratification, acceptance, accession or approval, which included countries such as Germany, Guinea-Bissau, Spain, Tunisia and Uzbekistan. A few NGOs, such as the WWF, have launched an initiative to accelerate the ratification process, but it is likely to be several years before the Convention is ratified. The campaign by the NGOs has brought additional attention to the issue of management of transboundary rivers and lakes. However, delays in the ratification of this Convention indicate two contributory factors: (a) management of transboundary water courses is not a priority issue in the world's political agenda, and (b) the countries that have transboundary rivers appear to prefer to have bilateral or multilateral negotiations between the co-basin countries, and do not appear to be in any special hurry to ratify the Convention.

Third, global interest in transboundary water management has been further heightened, because of the recent discord between the Nile Basin countries. The countries failed to agree on a treaty in May 2010, when the five upstream countries (Ethiopia, Kenya, Uganda, Rwanda and Tanzania) decided to sign an agreement without the basin heavyweights, Egypt and Sudan. Indeed, the transboundary issue had received increased international attention earlier when Pakistan decided to go straight to arbitration without considering the other options available under the Indus Water Treaty (Biswas, 1992). It is worth noting that the Indus Water Treaty has often been held as a showcase where the two signatories, India and Pakistan, had gone through two wars but the Treaty had functioned reasonably well during the past five decades. It illustrates the difficulties with static water treaties, since conditions change over time and the countries concerned find that treaties become increasingly out of tune with the new conditions and realities.

All these, and other associated, reasons have put the management of transboundary rivers and lakes in the international limelight. Yet, there are also other unjustifiable reasons why global attention has focused on this issue. Some water professionals, and also people with somewhat limited knowledge and appreciation of water issues, have repeatedly claimed, in recent years, that countries are likely to go to war with each other because of increasing water scarcity. The national and international media have given this idea of water wars considerable attention. These somewhat sensational claims have further increased global interest in the issue.

According to this linear and incremental thinking, the global demand for water is rapidly outpacing the supply available. As the world population increases, the demand for water would increase concomitantly to provide more food and energy, and to satisfy burgeoning domestic and industrial water requirements. With a simplistic but faulty reasoning and an extremely dubious database, both somewhat similar to *Limits of Growth* (Meadows *et al.*, 1972) discussions, several major international institutions have predicted that, by 2025, some two-thirds of the world's population would live in areas of moderate to serious water stress (WHO/UNICEF, 2005). Fortunately, such statements have no logical and scientific rationale. The demand for water, according to these institutions, is growing exponentially, but they assume erroneously that management practices, economic instruments and technology would grow only by discrete and limited amounts. They also erroneously assume water to be a finite resource, like oil or coal, which, once used, break down into various components and cannot be used again. The fact that water is a renewable resource, and, with good management practices, can be used, treated and re-used several times does not figure in this linear thinking.

With this simplistic thinking, many international institutions and water professionals predict that the world is facing an unprecedented water crisis, which will make many co-basin countries on the transboundary water courses go to war with each other. This has created a vicious circle: the more publicity these institutions and water professionals receive, the increasingly grim their claims of the world's water future become. Sadly, water crisis and water wars have become a growth industry. For example, if one searches for "water crisis" in Google, 12,400,000 citations may be found in the English language alone. Similarly, if "water conflicts" is searched, 10,500,000 citations would be found (these numbers correspond to those found at the time of writing this article). With such a wide coverage, the prevailing wisdom is that water scarcities will lead to conflicts, and even wars, between co-basin countries.

This thinking is incorrect. The fact is, water management practices in most of the countries of the world have been historically poor and continue to be poor. There is no question that if the present rate of inefficiency and complacency, in both developed and developing countries, continues in the future, the world would face a water crisis that would be unprecedented in human history, in terms of both quality and quantity. However, the potential threat of such a crisis and the increasing realization that many of the water problems can be solved by better management practices, including the use of good and enlightened economic instruments, institutional innovations and adoption of technological advances, have already started to create positive feedback loops, which are improving water-use patterns and efficiencies in many sectors in several countries. The current indications are that these positive developments will intensify in the future, which would greatly reduce the magnitude and intensity of the widely-perceived water threat.

As both the water professionals and the national and international institutions begin to realize that much of the world's water problems of the future can be resolved with the currently known, or available, management practices, technology and investment funds (Biswas & Seetharam, 2008), there is likely to be an increasing focus on selecting and using good management practices. In fact, there is some anecdotal evidence that this is already happening, albeit at a slower rate than desirable, in certain parts of the world. What is now needed is an accelerated drive to improve water management practices in most parts of the world, which would ensure that the discussions of water wars and water scarcities become increasingly irrelevant.

Implementation of better water management practices will have a profound effect on the management of transboundary rivers and lakes, and the changes in the mindsets of policy-makers and the general public in the co-basin countries. Instead of the current preoccupation with conflicts, both water professionals and policy-makers are likely to focus their attention on cooperation and collaboration between the countries, not only with respect to water but also in terms of a whole spectrum of development issues, such as agriculture, energy, industrial development, intermodal transportation (including navigation), which will invariably result in a very significant win-win situation for all the countries concerned (Ahmad *et al.*, 2001; Biswas & Uitto, 2001; Biswas, 2008a,b).

While much of the global attention of recent years has been focused on the wrong problem definitions, such as water wars and water conflicts, some countries are realizing that there is a much better solution. This will require implementation of good water governance to reduce demands, a search for unorthodox "out-of-the-box" solutions, intensive cooperation between the countries, and the use of water as an engine for economic development, poverty alleviation and environmental conservation. There are signs that this is already happening. However, while the mainstream water profession and the media have been preoccupied with water wars and conflicts in transboundary rivers, they are not aware of the good cases of cooperation and collaboration that have brought untold benefits to the people of the countries concerned.

This chapter focuses on the benefits of cooperation on transboundary rivers, as well as on the cost of non-cooperation between countries, with special emphasis on the South Asian countries.

Cooperation for Regional Development

In most Asian transboundary rivers, water allocation treaties between the relevant co-basin countries have been very difficult to negotiate. The Indus Water Treaty was unusual in the sense that the negotiation process, lasting just less than one decade, was short. However, it should be noted that Asia was a different continent half a century ago, and the World Bank, which acted as a facilitator to make this Treaty possible (Biswas, 1992), was held in very high esteem in the subcontinent at that time. In addition, the political leaderships in both India and Pakistan were enlightened, and the leaders of both countries truly wanted a solution. Equally, the relationships between the two countries were significantly more positive compared to what exists at present. Furthermore, Eugene Black, the then World Bank President, was willing to take the risk of possible failure in the negotiations, in contrast to the mostly risk-averse Presidents who have followed him. The "carrot" that the World Bank dangled in front of the two countries consisted of significant funding for projects in both, provided they reached a mutually acceptable agreement. This proved to be a very important practical incentive to expedite the Treaty.

These were some of the very special conditions that contributed to the success of the agreement on the Indus Water Treaty. The timing of the negotiations was most opportune. In the current situation, where the distrust between the two countries is high, and the importance and respect of the World Bank in Pakistan and especially India, is significantly lower than in the 1950s, it is highly unlikely that such a feat could now be duplicated. This cannot be seen as a positive development, since the earlier cooperation agreement has been of immense economic and social benefit to both countries.

In contrast, in recent years, in most Asian transboundary rivers, agreements have been difficult to negotiate between the appropriate co-basin countries because of many interrelated factors, among which are historical rivalries, political mistrust, asymmetric power relationships, increasing nationalism, short-term requirements of national political parties as compared to long-term national interests, growth of religious extremist groups, negotiations exclusively on water issues which invariably reduces water allocation to a zero-sum game, absence of properly formulated negotiating frameworks that could consider an overall development spectrum which could contribute to improving the standard of living in the countries concerned, emergence of other issues of conflict between the countries which adversely affect the negotiating atmosphere, and the presence of many vociferous, media-savvy and single issue NGOs that are more interested in promoting their own agendas and dogmas than improving the quality of life of the people whom they often claim to represent.

While the priority and importance of all these factors vary from one transboundary river of South Asia to another, individually and collectively they have seriously hampered cooperation in most of the transboundary rivers of the region, especially between Bangladesh and India, and India and Nepal.

Progress on managing transboundary rivers for mutual benefit has been mostly dismal due to non-water-related reasons, the deep-rooted mutual distrust, and sometimes even hostility. Accordingly, the benefits foregone by each of these three countries not using water as an engine for economic and regional development have been very substantial (Verghese 1990, 2007). This constitutes an appalling situation, especially when the extensive and abject poverty that exists in all three countries is considered. In fact, it is little known that a greater number of absolutely poor people now live in the

Ganges-Brahmaputra-Meghna (GBM) basin than in all the countries of sub-Saharan Africa combined. Considering the current level of poverty that exists in this region, none of these three countries can afford to continue with the unacceptably low level of cooperation which has greatly contributed to the sad situation.

Another important factor that is often lost in this debate is the fact that if developments of fossil fuels or mineral resources are delayed, these resources are not lost to the nations or to future generations. They remain in the ground, untouched, and can be exploited in the future whenever the countries decide to do so. The benefits will accrue whenever such resources are utilized. In contrast, if water is not used for hydropower generation or agricultural production at a given time, the potential benefits to the society are lost forever: they can never be recovered for societal benefit.

The only example in South Asia where cooperation on water-related developments has been the norm, rather than exception, has been the one between Bhutan and India. In fact, the GBM basin provides two excellent but contrasting examples of the very substantial benefits that can accrue when the countries concerned decide to collaborate actively for very substantial mutual benefits (India and Bhutan), and also equally the very substantial benefits that are foregone because the countries eschew the pursuit of common development goals for whatever reasons, some of which may be real but others could be imaginary (Bangladesh and India, and Nepal and India). Figure 1 shows the transboundary river systems of Bhutan, Nepal, India and Bangladesh (Biswas *et al.*, 2009). The benefits of cooperation and the cost of non-cooperation within the context of the GBM basin will be discussed next.

Fig. 1 Transboundary rivers of Bhutan, India, Nepal and Bangladesh (source: Biswas *et al.*, 2009. Copyright: Third World Centre for Water Management).

Bhutan and India: A Symbiotic Positive Relationship

In the area of transboundary water management, the constructive collaboration between Bhutan and India during the past three decades, which has brought very significant benefits to both countries, is basically unknown. Because of the size of Bhutan and its small population, the benefits to the country have had enormous impacts. In contrast, while benefits to India have also been significant, because of its size and population, they have not been a "game-changer" as has been the case for Bhutan. The experience definitively shows that, given enlightened leadership, political will and mutual trust and confidence, the benefits of cooperation in transboundary rivers often could be very substantial. Regrettably, however, in the present world, water conflicts attract considerably more attention than cooperation, and the proponents of conflicts receive far more attention relative to those who prefer cooperation. Thus, not surprisingly, the positive results of this very good collaboration between India and Bhutan are hardly known, even in the Indian subcontinent, let alone in the rest of the world. The collaboration between India and Bhutan is an excellent example of how transboundary rivers can be managed within an overall collaborative development framework which uses water developments as an engine for economic growth and poverty alleviation in a highly impoverished region.

Bhutan, often known as the Hermit Kingdom, was basically inaccessible to the world until 1960. When this landlocked country, located on the Himalayan mountain range, initiated its first development plan in 1961, it had by far the lowest per capita income in South Asia and one of the lowest in the developing world. Because of the mountainous nature of its terrain, its agricultural potential is limited. Its high mountainous location, however, provides the country with unique advantages, especially in terms of its hydropower development, which is estimated at 20,000 megawatts (MW), slightly less than one-quarter of the potential of its western neighbor, Nepal. However, in terms of population, Bhutan is much smaller than any of the other GBM basin countries.

Bhutan realized, sometime ago, that one of its main natural resources is water, and, if the country is to develop economically and make social progress, it must develop its water resources wisely and efficiently. Since nearly all of its water is transboundary in character, it decided to cooperate closely with India to develop these resources. Bhutan also recognized the following facts:

– Water development is not an end by itself, but only a means to an end, where the end is to improve the lifestyles of the people of the nation through a variety of complex interrelated socio-economic pathways.
– Alone, it cannot develop its water resources efficiently and quickly, because the country lacks investment capital and necessary technical and management expertise.
– Even if its water resources are developed, it will not be able to take full advantage of the resulting benefits exclusively within the national territory because of its small and dispersed population. In other words, the country simply does not have enough capacity to absorb all the benefits that could be generated by the water development activities.

Accordingly, Bhutan embarked upon a very different path, compared to either Bangladesh or Nepal, to develop its transboundary water bodies. It decided that the most efficient solution would be to develop its water resources in close collaboration with, and with the support of, its southern neighbor, India, with whom it shares its transboundary rivers.

Around 1980, Bhutan initiated a plan to develop the hydropower potential of the Wangchu Cascade at Chukha, in cooperation with India. Following extensive consultations, India agreed to construct a 336 MW run-of-the-river project at Chukha, on the basis of a 60% grant and 40% loan. The estimated cost of the project was INR 2,450 million. It was commissioned in stages from 1988 onwards. The project was so successful that it had covered its costs by 1993. The generating capacity was later increased to 370 MW. Because of the Indian support to the planning, construction and management of the project, Bhutan agreed to sell the excess electricity from the project, that it could not use, to India at a mutually agreed rate. A 220-kilovolt (kV) transmission line was constructed, which linked the Bhutanese capital, Thimpu, and the city of Phuntsholing on the Indian border, from where electricity was subsequently supplied to four Indian states.

The agreement between the two countries is that the electricity generated will be first used to satisfy Bhutan's own internal needs. Before the construction of the Chukha plant, electricity in Bhutan was generated by diesel and mini-hydro plants. Thus, the total electricity generated was extremely limited. Transporting diesel to a landlocked and mountainous country was an expensive and complex process; it was also inefficient. Not surprisingly, in 1980, per capita energy consumption in Bhutan was only 17 kWh, which was less than 10% of that of India, at 173 kWh.

Bhutan's per capita electricity consumption has steadily increased since the Chukha project became operational. For example, by 2008, Bhutan's per capita energy consumption, at 370 kg of oil equivalent, had already almost caught up with India (385 kg), was the same as in Pakistan (370 kg), and significantly higher than in Bangladesh (125 kg) or Nepal (47 kg).

The unit cost of hydropower generation has steadily declined since the Chukha plant was first constructed, because of greater and more economic scales of production and increasingly more efficient operation and management. The electrical network has steadily expanded to different parts of Bhutan, which has meant reduced use of fuel wood, and of diesel that had to be imported from India. Reduced fuelwood use has reduced the deforestation, and so has a beneficial impact on the forests and the environment.

The electricity produced in excess of the requirement of Bhutan is purchased and used by India as peak power through its eastern electricity grid. Initially, the two countries agreed to have two different pricing patterns for firm and secondary power. Later, the two tariffs were amalgamated into one, and, subsequently, the tariff that was initially paid by India was revised upwards four times. The revenue that Bhutan has been receiving from its electricity sales to India has not only serviced its debt load for the Chukha project without any problem, but also left enough surplus to finance other development activities, and to support some social services, including increasing the salaries of its civil servants. In addition, electricity has provided the impetus for Bhutan's industrialization and commercial development.

Since the construction of the Chukha project has proved to be beneficial to both the countries, they have agreed to expand their collaborative efforts to other new hydropower projects. Bhutan realized that the revenues from the development, use and export of its hydropower potential can accelerate the economic and social development of the country, and can contribute very significantly to poverty alleviation. The arrangement has also been beneficial to energy-hungry India, whose electricity requirements have been increasing in recent years at 7–9% per year. The decision for mutual collaboration on transboundary rivers has proved to be an important win-win situation for both countries.

India and Bhutan have subsequently collaborated with the funding and construction of a 45-MW run-of-the-river hydropower station at Kuri Chu. Similar collaborative efforts have taken place, or are under active consideration, for Chukha II (1020 MW) and Chukha III (900 MW, with a storage dam). In addition, the two countries signed an agreement in 1993 to study the feasibility of a large storage dam on the Sunkosh River. When all these projects are completed, and assuming the unit price paid by India for electricity will continue to be revised upwards periodically, Bhutan can easily earn more than US$100 million annually in the foreseeable future from the sale of hydropower to its neighbor. Considering that its present population is only a little over 2 million, this sale of hydropower to India represents a very substantial income for this relatively small country, that will accrue regularly, year after year. Because of this success, not surprisingly, Bhutan's development framework, Vision 2000, envisages careful and progressive utilization of its 20,000 MW hydropower potential as an important means to propel the country forward and upward so as to ensure a better quality of life for all its citizens.

The win-win approach used by Bhutan and India is a good example of how transboundary water bodies can be successfully managed by the basin-sharing countries for regional economic development, which can directly contribute to improvements in the quality of life of the people of both countries through income generation, poverty alleviation and environmental conservation.

Viewed from any direction, the collaboration between the two countries has been mutually very beneficial, including enhancement of regional peace and stability. These water-based developments have meant that Bhutan's per capita GDP has increased from being the lowest of any South Asian country in 1980, to being the highest by far in the GBM region at US$1,932.8 in 2008, compared to US$1,061.3 in India, US $1,010.2 in Pakistan, US$493.7 in Bangladesh and US$465.4 in Nepal. If the current trends continue, and it is highly likely that they will, by 2015, Bhutan would have the highest per capita GDP in the whole of South Asia, all due to its farsighted and enlightened approach to developing its transboundary rivers collaboratively with its neighbor.

Nepal, India and Bangladesh: A Missed Opportunity

In contrast to the successful case of Bhutan and India, the last 20 years have proved to be a missed opportunity for Nepal, India and Bangladesh because of continuing mistrust, and due to the "big brother–small brother syndrome". These developments illustrate the validity of the perceptive views of Jawaharlal Nehru, the first Prime Minister of independent India, who urged the people to override national conflicts. Nehru deplored the inability to overcome not only the "narrow boundaries of geography but, what is worse, of the minds".

The bilateral negotiations between Nepal and India, and India and Bangladesh have resulted in some agreements, and even treaties. However, real progress to use the waters of the river systems as a catalyst for economic development and poverty alleviation in the region has been minimal. Good historical analyses of the negotiations between Bangladesh and India can be found in Abbas (1982), and of those between all the countries in Verghese (1990), who also provides an excellent and objective analysis as to why the negotiations have failed to produce good results for all the countries concerned.

Had the three countries, Nepal, India and Bangladesh, jointly approached the planning and management of the transboundary rivers in a positive and constructive spirit, the benefits to all three, in terms of regional development, poverty alleviation and improvements in the quality of life of their people, most certainly would have been very substantial. Regrettably, this has not happened, partially because of political uncertainties that clouded the negotiations and partly because of asymmetric power relationships between the three countries. Many of these constraints should have been overcome by the Gujral doctrine of the mid-1990s, which very specifically eschewed absolute reciprocity in India's relationships with its smaller neighbors. While this new doctrine produced a burst of enthusiasm and activities between the three countries, this momentum could not be sustained for many different reasons. Accordingly, this was a missed opportunity for all the three countries. In retrospect, this perhaps hindered the progress and economic development of Bangladesh and Nepal more than India, since they had far fewer development options compared to India.

The overall situation of the region is not encouraging, since half of its population currently lives below the poverty line. In fact, in spite of recent economic advances, the total number of poor people in this region has continued to increase. Not surprisingly, the various health and social indicators for the countries still leave much to be desired.

Water is one of the few resources this region has that can promote long-term economic development. The countries need to formulate and implement cooperative strategies and joint action plans in which water could act as the catalyst for economic take-off. Several options and opportunities for collaborative efforts have existed for decades in areas such as hydropower generation, flood management, drought mitigation and agricultural development. However, progress has been very slow.

The GBM region is characterized by endemic poverty (Rahman, 2009). The performance of the region with respect to social indicators, such as economic growth, education and health, is disappointing, especially in comparison with other regions of the world. About 40% of the developing world's poor people (with a daily calorie intake of less than 2200–2400 kcal) live in this region; and, even though there has been a decline in poverty in recent years, the absolute number of poor people has increased due to population growth. Adult illiteracy is still very high, especially among women. The three countries spend a lower share of public expenditure on education than the world average.

Health indicators are also dismal in the region. Infant (under 1 year) and child (under 5 years) mortality rates in these countries are much higher than those of many other developing countries, as well as the world average. Although access to clean water has improved in recent years, only a limited population has proper access to wastewater collection and treatment.

Nearly 45% of the land of the GBM region is arable, but per capita availability of arable land is very small, around one-tenth of a hectare, which is almost half the global average. One other crucial element to be taken into consideration in envisioning a sustainable development framework for the GBM region is the trend in urbanization. In Bangladesh, India and Nepal, annual urban growth rates (1995–2000) were 5.2%, 3.0% and 6.5%, respectively. These rates are much higher than those of Europe (0.5%), Latin America (2.3%), Australia (1.2%), the USA and Canada (1.2%) and Japan (0.4%). While the proportions of urban population in the three GBM countries are 20%, 27% and 14% respectively, they are expected to rise to over 50% in the case of India and Bangladesh, and to about 22% for Nepal by 2025. This change in the spatial distribution and localization of population will have significant implications for water, energy and other related demands for natural resources and their socio-environmental impacts.

Despite the poor socio-economic status of the region, it has rich natural endowments of water, land and energy. It is indeed an agonizing paradox. The development and utilization of these natural resources in an efficient manner have never been sought by the countries due to past perceived differences, a legacy of mistrust, lack of goodwill and an absence of sustained political will (a very important factor for development), which could lift millions of people out of the poverty trap. The abundance of water in the GBM region, as a shared resource, could be a principal driver of economic development for the millions of poor people. The shared river systems can be optimally developed only through collaborative efforts. It is imperative, therefore, to formulate a framework for the sustainable development of this region in a long-term time frame and on a cooperative basis, which would be acceptable to the three countries and implemented.

Concluding Remarks

The framework for sustainable development in South Asia should be based on a vision of poverty eradication and sustained improvement in the living conditions of many hundreds of millions of its inhabitants (Biswas & Uitto, 2001). The world's largest concentration of economic misery is to be found in this region. There is no reason for such abject poverty, given the rich bounty of its natural resources, especially water, waiting to be harnessed.

However, a lack of trust between the countries and the absence of forward thinking have consistently bedevilled the relationship among the co-riparians for nearly half a century, and compounded the poverty and deprivation in the region. This pernicious mindset has eroded goodwill and confidence, and has generated mutual mistrust and suspicion. The situation is further compounded by the failure of the political leaders to create public opinion in favor of formulating and implementing a vision for regional cooperation and development.

The drivers that would influence the conditions towards achieving the regional vision include population growth, urbanization, technology, globalization, governance and environment. Technological changes, manifested through innovation/adoption of new products and techniques, can enrich human resources through capacity development. The South Asia region might benefit from transferring water-related technology from industrialized countries as well as from within the region, especially concerning irrigation efficiency, pollution control, water storage, disaster management and information systems. The contemporary process of globalization could be another driver in the region's long-term vision for sustainable development. The region would benefit from trade liberalization, greater capital mobility and technology transfer; but, at the same time, it is important to be vigilant against potential instability and the risk of greater inequality in income distribution. To address this issue effectively, it is necessary to establish good governance at all levels of society, reflected in accountability, rule of law, elimination of corruption and participatory approaches (Biswas & Tortajada, 2010).

The regional vision formulation can be approached under three scenarios: pessimistic, optimistic and plausible. A scenario is a possible course of events. The pessimistic scenario is basically a business-as-usual approach under the assumption of a *status quo* and "do nothing" response strategy; this approach is unsustainable and unacceptable for the long term. The optimistic scenario is the other extreme, which is overly ambitious, utopian and an unrealistic goal to pursue. In between lies the plausible scenario. It is pragmatic to seek sustainable water resource management for the region through genuine cooperation and collaboration, as has been the case between Bhutan and India.

The experience from the South Asian countries clearly indicates that, over a longer time frame, the countries have no other alternatives but to cooperate with each other in managing their transboundary rivers. In the entire human history, no two countries have gone to war because of water. While water wars may be of interest to the media, it can be safely predicted that if there were ever to be a war between the countries sharing a transboundary river, the root causes would be non-water reasons. Water would, at best, be one of the many tertiary causes, but never the primary or secondary reason.

The benefits of cooperation can be seen by the results of the India–Bhutan relationship, while the costs of non-cooperation can be seen by the Nepal–India–Bangladesh experience. In the final analysis, the costs of non-cooperation in the South Asian transboundary rivers will be paid not by the politicians and the media, but by the hundreds of millions of poor people, the vast majority of whom would be forced to live in abject poverty for decades to come.

References

Abbas, B.M. 1982. *Ganges water dispute*. Dhaka: University Press.

Ahmad, Q.K., Biswas, A.K., Rangachari, R., and Sainju, M.M. 2001. *Ganges Brahmaputra-Meghna region: a framework for sustainable development*. Dhaka: University Press.

Biswas, A.K. 1992. Indus Water Treaty: the negotiating process. *Water International*, 17 (4), 201–209.

Biswas, A.K. 2008a. Management of transboundary waters: an overview. *In*: O. Varis, C. Tortajada, and A.K. Biswas, eds. *Management of transboundary rivers and lakes*. Berlin: Springer, 1–20.

Biswas, A.K. 2008b. Management of Ganges-Brahmaputra-Meghna systems. *In*: O. Varis, C. Tortajada, and A.K. Biswas, eds. *Management of transboundary rivers and lakes*. Berlin: Springer, 143–164.

Biswas, A.K., and Seetharam, K.E. 2008. Achieving water security for Asia. *International Journal of Water Resources Development*, 24 (1), 145–176.

Biswas, A.K. and Tortajada, C. 2010. Future water governance: problems and perspectives. *International Journal of Water Resources Development*, 26 (2), 129–139.

Biswas, A.K. and Uitto, J.I. 2001. *Sustainable development of Ganges-Brahmaputra-Meghna Basins*. Tokyo: United Nations University Press.

Biswas, A.K., Rangachari, R., and Tortajada, C. 2009. *Water Resources of the Indian Subcontinent*. New Delhi: Oxford University Press.

Meadows, D.H., Meadows, D.L., Randers, J., and Behrens, W.W. 1972. *The limits to growth*. New York: Universe Books.

Rahman, M.M. 2009. Principles of transboundary water resources management and Ganges treaties. *International Journal of Water Resources Development*, 25 (1), 159–173.

Verghese, B.G. 1990. *Waters of hope: Himalaya-Ganga development and cooperation for a billion people.* New Delhi: Oxford and IBH Publishing.

Verghese, B.G. 2007. *Waters of Hope: Facing New Challenges in Himalaya-Ganga Cooperation.* New Delhi, India: Research Press.

WHO/UNICEF, 2005. *Water for life: making it happen.* Geneva: World Health Organization, WHO Press.

The Singapore–Malaysia Water Relationship: An Analysis of the Media Perspectives

CECILIA TORTAJADA[1,2] & KIMBERLY POBRE[2]

[1]International Centre for Water and Environment (CIAMA), Zaragoza, Spain
[2]Lee Kuan Yew School of Public Policy, National University of Singapore, 259772 Singapore

ABSTRACT   *This paper explores the role of the media in the Singapore–Malaysia water relationships, focusing on the water negotiations during the 1997–2004 period. Detailed examination of reports from the print media of Singapore, Malaysia and international sources constitutes the basis for a discussion of the roles the media have played between the two countries. The analysis shows how the media slowly evolved from being mainly a reporter to becoming an active platform for communication between the interested parties, acting both officially and unofficially, as well as directly and subtly, leading to shaping public opinion in either of the two countries, particularly regarding their water relationship.*

Introduction

Singapore is a city-state of 4.987 million inhabitants with a land area of 710.3 km (Department of Statistics, Ministry of Trade & Industry, 2010). With a rapid rate of economic growth, an open-door policy for foreign investment, and clear priorities on economics for major decisions related to its development, Singapore is currently one of the most developed countries in Asia with a per capita GDP second only to Japan (World Bank, 2010). Malaysia, located north of Singapore, has a population of 28.31 million people and an area of 330,803 km (Department of Statistics, 2010). With a robust economic recovery process underway, the country's medium-term growth outlook focuses on the implementation of structural reforms that will unleash its innovation potential, shifting the sources of comparative advantage from low costs to high value.

Singapore and Malaysia have established unique bilateral relations. Their geography, historical heritage, economy and culture have contributed to an intricate and closely interdependent relationship. Both the countries, together with Sabah and Sarawak, were once united as the Federation of Malaysia in 1963. The merger soon failed and Singapore became independent in 1965. From that date on, Singapore's lack of natural resources to support its economic growth and social development, mainly in terms of water resources, made the leadership of the country aware of the importance of developing and implementing clear visions, long-term planning and forward-looking policies and strategies that would provide it with enough flexibility to achieve its increasingly ambitious development plans (Ghesquière, 2007; Yap *et al.*, 2010).

13

With very clear objectives in mind, national water strategies in the city-state have comprised the formulation and implementation of innovative policies, increasingly more efficient water management practices, well-planned development of local resources, heavy investments in desalination and extensive reuse of wastewater (Tortajada, 2006; Ching, 2010; Lee, 2010; Luan, 2010). Additionally, for decades, long-term water conservation practices have focused on catchment management practices, mixes of economic instruments and very well-targeted education programmes that have been adapted to the present and foreseen future needs of the city-state.

A historically important source of water for Singapore has been imported water from Johor*, Malaysia, which will last, at least, until 2061. Four water agreements have been signed with that purpose: in 1927 (no longer in force), 1961, 1962 and 1990, allowing Singapore to import water from Johor and allowing Johor to buy treated water from Singapore (see Appendix A). These agreements illustrate a history of consistent cooperation between the two countries on the issue of water, even when their "water relationship" has not been exempt from serious disagreements and differences in opinion at various times in history (see Kog, 2001; Long, 2001; Kog et al., 2002; Lee, 2003, 2005, 2010; MICA, 2003; NEAC, 2003; Chang et al., 2005; Saw & Kesavapany, 2006; Sidhu, 2006; Dhillon, 2009; Shiraishi, 2009; Luan, 2010).

Throughout the years, an important player in shaping the water relationship between the two countries has been the media. The media has acted both as a reporter and as a vehicle of communication, both officially and unofficially, to its own public but also to the interested parties in the other country. At some point, the media has been described as provocative on both sides (MICA, 2003; Chang et al., 2005), contributing to "heighten emotions" in both countries (Chang et al., 2005, p. 3).

The fundamental importance of the media in regard to the water relationship between the two countries lies in the fact that, except for very few primary documents publicly available, such as those released by the Government of Singapore in 2003 (MICA, 2003), public information on the Singapore–Malaysia water relationship has been available primarily through the media. A clear indication is that the overall studies on this topic rely, sometimes heavily, on information disseminated initially in media reports.

The objective of this chapter is thus to analyse the perspectives and roles of the print media in the Singapore–Malaysia water relationship, mainly at the time when bilateral water negotiations were extensively covered (1997–2004). The management of the media in Singapore and Malaysia by the respective governments is beyond the scope of this chapter, and has been analysed extensively elsewhere (e.g. Ang 2002, 2007; George 2007; Kenyon 2007; Kim 2001).

The sources consulted for this analysis are from Singapore and Malaysia, as well as international. These sources are as follows:

- Singapore media: *Berita Harian* (Malay); *Berita Minggu* (Malay); *The Business Times; Nanyang Siang Pau* (Chinese) and *The Straits Times.*
- Malaysia media: *Bernama; Mingguan Malaysia* (Malay); *New Straits Times; Oriental Daily News* (Chinese); *Sin Chew Daily* (Chinese); *The Star and Utusan Malaysia* (Malay).
- International media: *BBC Summary of World Broadcasts; BBC Monitoring; Birmingham Post; Business World; Deutsche Presse-Agentur; International Herald*

Tribune; Newsweek; The Economist; The Independent; The Korean Herald; The Nikkei Weekly; Xinhua; Waste News.

Before analysing the Singapore–Malaysia water relationship as viewed by the media, a general overview of the media industries in both countries is presented.

The Media in Singapore and Malaysia

The media industries in both Singapore and Malaysia are highly regulated by the respective governments. Characterised by numerous brands, but owned by a few companies, media content may not differ significantly from one medium to another. Both media structures are also described to have pro-government tendencies, which might have implications on the industry's attitude and reporting. While, ideally, coverage on the water relationship in both countries should be impartial, objective and factual, in real terms, reporting reflects to a significant extent the views held in their respective countries, a reason why portrayal of the events related to water negotiations may not necessarily be the same.

Singapore Media

There are effectively only two print media companies in Singapore: Singapore Press Holdings (SPH) and MediaCorp. The SPH is a predominantly print media company, while MediaCorp, though mainly a broadcasting company, has one newspaper in circulation. The SPH has a 40% ownership of the MediaCorp Press, and MediaCorp Press owns a substantial stake in SPH. Even though both companies are private groups, their management is linked to the government, generally holding a government-favourable stance (Ang, 2002; Tan, 2010).

Media market ownership in Singapore is often described as monopolistic (Ang, 2007; Tan, 2010). Except for *Today*, which is owned by MediaCorp Press, all print media is owned by SPH. The SPH publishes 17 newspapers in four languages and has 77% of the readership in Singapore above 15 years old. The several newspapers are as follows: *The Straits Times, The Sunday Times, The Business Times, The Business Times Weekend, The New Paper* and *The New Paper on Sunday* (in English); *Lianhe Zaobao, Zaobao Sunday, Lianhe Wanbao, Shin Min Daily News, zbComma* and *Thumbs Up* (in Chinese); *My Paper* (English and Chinese); *Berita Harian* and *Berita Minggu* (Malay); and *Tamil Murasu* (in Tamil). *The Straits Times* is considered to be the most influential English newspaper in Singapore. In August 2010, it was the newspaper with the largest circulation, with 365,800 copies distributed per day (SPH, 2010).

The broadcasting media is dominated by Media-Corp, owned by Temasek Holdings, the government's investment arm. Internet-related media in Singapore is less restricted than print media, but is subject to controversial licensing regulations (Ang, 2007).

Media regulation in Singapore is extensive. The Newspaper and Printing Presses Act (Government of Singapore, 1974) requires publishers to renew licenses yearly. Media companies are also required by law to be public entities with no single shareholder controlling 12% or more of a newspaper company without first obtaining government approval. Furthermore, the government has the legal authority to approve ownership and transfer of management shares that hold higher voting power in these companies. Publication that is "contrary to the public interest" is prohibited under the Undesirable

Publications Act (http://statutes.agc.gov.sg/non_version/cgi-bin/cgi_retrieve.pl?&act-no=Reved-338&date=latest&method=part).

Malaysian Media

In Malaysia, the United Malays National Organisation (UMNO) not only is the leading political party but also owns most of the main newspapers. Print media in Malaysia is in English, Malay, Mandarin and Tamil. The major companies are the New Straits Times Press (NSTP) and the Utusan Melayu Press (UMP). The NSTP publishes English newspapers, such as: *New Straits Times, New Sunday Times, The Business Times*; as well as *Berita Harian, Berita Minggu, Harian Metro* and *Metro Ahad*. The Malay-based press holding UMP circulates *Utusan Melayu, Utusan Malaysia, Mingguan Malaysia* and *Utusan Zaman*. Besides UMNO, other political parties, such as the Malaysian Indian Congress (MIC) and the Malaysian Chinese Association (MCA), are closely linked, in terms of ownership, to the media (Kim, 2001). Similar to Singapore, although there are numerous newspapers available in Malaysia, these are controlled by a few companies that are connected to the ruling party coalition (Shriver, 2003). In general, mainstream print media is not critical of the government (Kenyon, 2007).

Radio and television are also government-owned and controlled. The Internet, though still subject to some form of control, is the least restricted type of communication channel. However, the government has banned several websites and is known to employ a surveillance system that restricts access (Kim, 2001). Through the Communication and Multimedia Act, licensing for Internet providers is also required and provides for legal actions against content that is considered to be defamatory and false.

The Internal Security Act and the Printing Presses and Publications Act give mandate to the government in controlling the media. The Internal Security Act restricts coverage of matters that are considered a threat to national interest and security (Kenyon, 2007). The Printing Press and Publications Act stipulates both granting and withdrawal of media licenses. Under this bill, where media publishers have to renew licenses yearly, the government has the discretion to withdraw licenses without any obligation for explanation (Kim, 2001).

Singapore–Malaysian Water Relations as Viewed by the Media

Before presenting the perspectives of the media, it is important to introduce the situation regarding water negotiations during the 1997–2004 period, as addressed by the Ministry of Information, Communications and the Arts of Singapore (MICA, 2003) (see also Appendix B for a chronology of developments):

> The story began in 1998. Crises, they say, bring people together; so it was that at the height of the Asian financial crisis, the two countries began negotiations on a "framework of wider cooperation". Malaysia wanted financial loans to support its currency. To enable it to carry its domestic ground when acceding to the request, Singapore suggested that Malaysia give its assurance for a long-term supply of water to the Republic. Malaysia eventually had no need for the loans, and so negotiations turned to other matters of mutual interest. In particular, Malaysia wanted joint development of more land parcels in Singapore in return for relocating its railway station away from the current site at Tanjong Pagar.

Over the ensuing three years, more items were bundled into the negotiation package. Singapore added one request: resumption of its use of Malaysian airspace for military transit and training. Malaysia added three more: replacing the Causeway with a bridge, early withdrawal of the Central Provident Fund savings for West Malaysians working in Singapore and a higher price for the water it presently sells to Singapore.

Officials met; leaders corresponded; Singapore's Prime Minister Goh Chok Tong and Senior Minister Lee Kuan Yew took pains to visit Malaysian leaders at the capital, Kuala Lumpur . . . .

The water negotiations came to the forefront of the Singapore–Malaysia cooperation framework as early as 1995, and in connection with talks on the Malayan Railway and Malaysia's plans to invest in an electric train that would connect both countries (*The Business Times* (BT), 6 June 1997). By 1996, Malaysia's willingness to supply water to Singapore, contingent upon its domestic needs, was publicly announced (BT, 6 June 1997). This was reinforced in the two-day visit of Singapore's Prime Minister Goh to Malaysia in February 1998, where both countries released a joint communiqué re-affirming this (NST, 4 April and BT, 10 April 1998). The details of the new agreement were set to be refined within a 60-day period after the official visit. However, this deadline was not met because both parties could not reach an accord on the details (BT, 30 June 1998).

In 1998, the water negotiations were linked to a financial recovery package between Singapore and Malaysia that the latter subsequently passed in favour of a package approach for outstanding bilateral issues (BT, 18 December 1998). This package linked the issue of water to the negotiation of other bilateral issues including: "the Malayan Railway land in Singapore, the relocation of the immigration checkpoint for the Malayan Railway, the use of Malaysian airspace by Singapore aircraft, the transfer of Malaysian shares no longer traded on CLOB International to the Kuala Lumpur Stock Exchange, and the early release of CPF (Central Provident Fund) savings of Malaysian workers" (BT, 18 December 1998). There was very little media coverage reported on the status of the negotiations at that moment. Around the same time, Singapore conducted feasibility studies to source water from Indonesia (BT, 15 June 2000 and 21 Aug. 1998; *Straits Times* (ST), 16 Jan. 1999).

In September 2001, Singapore Senior Minister Lee Kuan Yew and Malaysian Prime Minister Mahathir reached an in-principle agreement to resolve a host of outstanding bilateral issues, including water (BT and ST, 5 September 2001). New to this agreement was the replacement of the Causeway bridge in favour of a railway tunnel that would connect both countries (ST, 5 September 2001). Official proposals and counter-proposals were exchanged between the two countries, followed by a new round of bilateral talks in July, September and November 2002. Several factors, such as delinking water from the package deal, the price of water and Malaysia's right to review the price of water, stalled the process. The package deal approach was subsequently dropped for an individual approach by the third round of talks in November 2002. However, no deal was reached. Malaysia only wanted to discuss the current price of water, while Singapore also wanted the issue of future water supply included in the agenda (ST, 21 November 2002). The impasse led to discussions that circulated in the press about seeking legal resource to resolve whether Malaysia had a right for a price review (ST, 21 November 2002).

In January 2003, Singapore released official letters from the negotiations in an effort to clear matters (MFAS, 2003). Several of these letters were published in Singapore's *Straits Times* (ST, 26 January 2003a,b; 28 January 2003) and also were made available on the website of the Ministry of Foreign Affairs (MFAS, 2003). This decision was heavily criticised by Malaysia (*New Straits Times* (NST), 28 January 2003), who released, in July 2003, a week-long series of advertisements on the dispute titled "Water: The Singapore–Malaysia Dispute: The Facts" (NEAC, 2003; NST, 13 July 2003).

Soon after that time, the official negotiations stopped. Malaysia stated that it would still honour current agreements but that negotiations were over (ST, 2 August 2003). In contrast, Singapore expressed intentions of letting the first of the water agreements expire in 2011 (NST, 6 August 2002).

A host of additional factors also affected the bilateral water relations. These included the quest of both countries to be independent from each other in terms of water: Singapore wanting water self-sufficiency and Johor's aims to be independent from Singapore in regards to water treatment; the possibility for Singapore to source water from Indonesia; the development of the water industry in Singapore supported by the production of NEWater[1] and desalination plants; and reclamation efforts in Singapore which were affecting Malaysia, and the dispute over Pulau Batu Putih/Pedra Branca.

Coverage of Water Relations

When analysing the role of the media in the Singapore–Malaysia water relationship, it is fundamental to consider the prevailing historical and political context between the two countries at specific times. Regarding bilateral issues, experience has shown that media coverage, in general, does not necessarily follow the philosophy of the media as a public sphere where "citizens discuss and deliberate matters of common interest and public concern, and hold the state accountable" (George, 2007, p. 94). In bilateral issues, the media tend to be rather nationalistic with homogeneous views which focus primarily on the interests of their own countries and reflect the views of the respective states. The case of the Singapore and Malaysia media was no different.

The coverage of the water negotiations across the two countries evolved over time. At first, coverage was dominated by news articles portraying positive images of cooperation during the bilateral negotiations, and frames were consistent across Singaporean and Malaysian news. Later on, however, coverage grew increasingly negative and framing of the issues soon differed. The local interest grew with the proliferation of opinions and editorial articles alongside the news articles. Bilateral negotiations soon became a domestic political issue, and the leaders used the media to clarify and explain the status of negotiations to their respective public.

It is important to mention, at this point, that mass media coverage of political issues is normally, and necessarily, selective. The media depends on frames to give coherence to relatively brief treatments of complex issues through selective views, choosing to highlight certain items and ignore others, frequently looking for consensus or disagreement on certain issues. Since the number and variety of issues that an audience can appreciate on specific subjects is normally limited, public debate is constrained on matters even when they are relevant, simply because citizens tend to focus on specific issues when constructing their opinion. In the case of the Singapore–Malaysia water relationship—as one would expect to be the case in any normal bilateral relations—the media could decide to transmit mainly the views of the own establishment with a clear objective to form a favourable public opinion.

Jürgen Habermas, the German sociologist and philosopher, argued that the function of the media has been transformed from facilitating rational discourse and debate in the public sphere, into shaping, constructing and limiting public discourse to the themes approved by media groups (Kellner, 2000). Nevertheless, one could equally argue that citizens trust their elites on many occasions, and, in this case, also their national mass media, on bilateral issues, because they perceive them to be credible sources of information. In addition, as discussed by Andina-Díaz (2007, p. 66), the influence of the media is "neither as significant as it was first thought to be, nor as minimal as was subsequently assumed". This is because citizenry does not necessarily follow the media blindly or accept passively all views as presented, since people have their own motivations and own biases which may or may not coincide with the views presented by the media. Consequently, as important as the media is to shape public opinion, and as consistent as the messages of the media can be with their respective establishments looking to shape public opinion, a fundamental player in the equation is the readership. Citizens are not necessarily mere spectators who allow the media to mould their opinion: they normally make use of their own perspectives and expose themselves to material with which they normally agree, often interpreting the media content to reinforce their own views and perspectives.

For this analysis, the period covered (1997–2004) for the bilateral negotiations, has been divided into three parts following the media's evolving role (Table 1). The media's traditional role as a reporter of events is clear in the early phase of the water negotiations that is covered by the first period. The second and third periods include the increased role of the media as a communication platform. Nonetheless, while the second period demonstrates the media's role for clarification of facts for interested groups other than officials, the third period is identified by the media playing the role of an unofficial medium for communication between the governments of Singapore and Malaysia.

Each of these periods is discussed in detail in the following sections. This includes media portrayal from different news sources as well as the consistency of both *media content* (consistency of news content regardless of media source) and *media framing* of the news (assessment of the portrayal of the media content between different sources where the same media content may be presented or highlighted in different ways defining and constructing a political issue or a public controversy). This analysis also includes classifying the data as positive, negative or neutral, and making a comparison between the three different media sources.

The Media as a Reporter

During the early phase of the negotiations (1997–1998), the media played largely a reporting role wherein news articles gave an account of the proceedings of the water negotiations between the two countries. Only the offices of the Prime Minister and the Foreign Minister of each country were directly involved in the negotiation process. In addition, there was a consistency in terms of the *content* that the Singapore and Malaysian media covered. This period mainly featured positive news coverage on the possibility of further cooperation on bilateral water agreements. Highlighted in these years were the countries' signing points of agreement (BT, 6 June 1997) with Singapore

Table 1 The role of the media in the Singapore–Malaysia water relationship, 1997–2004.

Media role	Period	Media content		Composition of print media	Media framing
		Bilateral news			Comparative media coverage
As reporter	First 1997–1998	Mostly positive content		News articles	Consistent
As reporter and platform for communication	Second 1999–2001	Increasingly negative content		News articles, opinions, editorials, forums and letters	Consistent with slight framing of news
	Third 2002–2004	Mostly negative content		News articles, opinions, editorials, forums, letters and pamphlets	Water negotiations were framed differently

willing to buy water from Malaysia and the latter willing to supply water to Singapore beyond 2061 (NST, 4 April 1998 and BT, 4 August 1998). By mid-1998, there were some negative news items on the stalled talks due to a failure to reach an agreement between the countries, and a change from linking the water deal to a financial assistance package to a host of other bilateral issues (BT, 18 December 1998). In spite of this, news coverage was still consistent in terms of content between Singapore and Malaysian and international media sources.

During this first period, media framing did not differ significantly between the two countries. However, there was a different choice of words between the Singaporean and Malaysian media with the same content being portrayed in different lights. For example, Singapore news reported that the UMNO youth "calls for suspension of talks" (BT, 4 August 1998), while, according to the Malaysian news, the UMNO youth "urges the Malaysian government to be firm" in dealing with Singapore (NST, 4 August 1998). Another difference was when Singapore news reported that Malaysia had "agreed" to continue to supply water to Singapore (BT, 6 November 1998), while Malaysian news stated that it would "not cut off water supply to Singapore" (NST, 5 August 1998). Regarding the loan assistance, Singapore news reported that Malaysia "sought Singapore's help" (BT, 6 November 1998), which it eventually chose not to take (BT, 18 December 1998), while Malaysian news reported that it had told Singapore that it "does not need the funds offered by Singapore" (NST, 18 December 1998). Thus, even though media contents in the two countries were fairly similar, the framing of the issues did differ across the two countries.

For these years, although there was consensus among the leaders to cooperate, there was no policy certainty, because no agreement could be reached. It is, thus, not surprising that in the first period, the media in both countries mainly played a role as informative reporter.

The Media as a Medium for Communication

The media's role changed as it increasingly became a medium of communication between the two countries. During the second period (1999–2001), the media's role as a medium of communication was mainly for interested groups in the negotiation process other than official representatives. Finally, throughout the third period (2002–2004), the media's role further expanded when it served as an unofficial medium of communication between the two governments on the water negotiations. These aspects are discussed below.

The Media as a Medium of Communication for Stakeholders other than Official Representatives

During the second period, 1999–2001, the media played a second role. From a media coverage that was dominated by news articles, this period saw an increase in opinion and editorial articles that became gradually more negative in both countries.

There was still a general consistency in terms of media content in Singapore and Malaysia. The nature of the media coverage extended beyond just reporting, becoming a medium for communication and clarification. The media played a role in addressing concerns raised by both sides. For instance, Malaysian opinion articles portrayed the unfairness of the situation and how Singapore was benefiting from the existing water deal (NST, 23 February 1999). Singapore opinion articles in turn portrayed how Singapore was not profiting from the deal while Malaysia was (ST, 3 March 1999 and NST, 17 June 1999). In addition, Johor's own interest in meeting its own water needs before Singapore's were also raised in the Malaysian Press (NST, 7 and 8 June 1999). This was then addressed by the Singapore media, reiterating that the country's water demand was contingent upon Malaysia satisfying its own needs first (BT, 11 June 1999). Thus, the media served as a platform for communication and clarification for the population of both countries.

Around this period, media coverage also voiced concerns on the over-dependence of Singapore and Malaysia on each other (ST, 2 December 2000, and BT, 24 April 2001). This might have contributed to other initiatives at this stage, where Singapore floated the idea of starting a partnership with Indonesia as an alternative source of water supply (BT, 15 June 2000, 15 February 2001 and ST, 2 July 2000a, 2 July 2000b), and of investing in more desalination plants (ST, 15 and 22 March 2001). It was also reported that Malaysia had decided to build a water treatment plant in Johor to reduce reliance on receiving treated water from Singapore (BT, 19 August 2000).

Given these events, media content was fairly consistent across the two countries and, even though it served to clarify and answer concerns raised by both sides, it covered roughly the same topics. Nevertheless, by then, the number of negative articles started to increase and the prolonged negotiations, which did not result in any agreement, may have facilitated this increased role of the media as an avenue for clarification. To some extent, this was to be expected given the increasing interest of the public in both countries on the status of the water negotiations. Compared to the first period, when only the official parties relayed information to the press, this interval saw various groups, such as Malaysia's UMNO youth group and Singapore's Worker's Party, expressing their thoughts and views on the water negotiations (NST and BT, 4 August 1998; ST, 7 September 2001).

The exchanges between the Singaporean and Malaysian media, which were absent in the coverage of the international media, show the beginning of the framing of the water negotiations. On the one hand, Malaysia started portraying how the past water deals were in favour of Singapore (NST, 8 June 1999) and how Singapore was profiting from them (NST, 10 and 17 January 2001). On the other hand, Singapore started to portray how both countries had benefited from the agreements (NST, 17 June 1999) and how it was not profiting from the water deals (ST, 3 March 1999; NST, 17 June 1999). Thus, within this role as a platform for communication and clarification, the media became a conduit to address the other country's concerns.

Though this period still projected an overall consensus among the leaders to cooperate, the degree of agreement began to change. For example, while there was a general consensus among the leaders, media coverage started to depict diverging and negative opinions. The prolonged negotiations were also an indication of heightened policy uncertainty. The media gradually became more focused on clarifying the respective government's views on various issues related to the negotiations, compared to its earlier role as a one-way (media to the public) reporter.

The Media as an Unofficial Medium of Communications

In contrast to the 1997–1998 and 1999–2001 periods, when communication and clarification were the primary roles of the media, during the third period, from 2002 to

2004, the media's role changed even further, when it started to serve as an unofficial medium for communication between the governments of Singapore and Malaysia. Furthermore, a distinct framing of the news was also more evident, where negative views were expressed on the Singapore–Malaysia bilateral ties. There was now a significant increase in the editorials, opinions and letters on the water negotiations.

The analysis below illustrates how the media played an unofficial role as a medium for communication between the two governments, and how the Singaporean and Malaysian media framed the news.

(a) Unofficial Medium for Communication between Governments

In one example, Foreign Minister Syed Hamid of Malaysia released news to the media about seeking legal recourse and cancelling negotiations with Singapore as early as October 2002 (BT, 25 October 2002; ST, 25 October 2002; NST, 14 October and 30 November 2002). However, no official notice was forwarded to Singapore, with the Foreign Minister indicating that the newspapers knew where Malaysia stood in seeking legal recourse (ST, 3 December 2002). Another instance was the agenda of the negotiations on whether it would include both current and future water abstraction by Singapore or only current water requirements. Malaysia stated that it would only discuss the price of current water (ST, 20 November 2002), while Singapore stated that both current and future water must be on the agenda (ST, 21 November 2002). Following the media reports, Singapore sought official clarification from Malaysia's Foreign Ministry to confirm the status of future water talks (ST, 1 December 2002). These examples illustrate how Singapore and Malaysia utilised the media as a platform for unofficial communication with each other.

In addition, the media played a role in delivering subtle signals and messages between the two governments. These signals captured the sentiments of both sides, which might or might not have come out during the official negotiations. For instance, when the Singapore media featured consecutive articles on NEWater's safety and quality (ST, 12 July 2002; BT, 12, 17, 26 July 2002), Deputy Prime Minister Lee talked about the possibility of NEWater replacing the water imported from Johor (ST, 13 July 2002), and Foreign Minister Jayakumar announced that water would no longer be a strategic vulnerability for Singapore (ST, 29 July 2002). On the other side, Malaysian media also delivered subtle signals to Singapore when Foreign Minister Syed Hamid said that NEWater would not affect Malaysia's stand in the negotiations (The Star, 18 August 2002), and Prime Minister Mahathir informed the media that Singapore was free to stop buying water from Malaysia (NST, 7 August 2002). Indirect signals, such as Johor officials negating the relevance of NEWater to Singapore's water security (ST, 11 August 2002), and a Malaysian MP's suggestion of selling sewage water to Singapore instead (ST, 3 October 2002), signalled Malaysia's position in the negotiations amid changing water dynamics.

Therefore, the media played an unofficial platform not only for communication on bilateral negotiation topics, such as the agenda and arbitration, but also for sending subtle signals to both countries. With the growing interest in the water negotiations, the media also served as a platform for both parties to explain and educate their domestic constituencies about the progress of the negotiations. The speech of Singapore's Foreign Minister Jayakumar to the Parliament detailing the water talks was published in the newspapers (ST, 26 January 2003c). Simultaneously, several of the letters from which the Foreign Minister quoted extensively were released to *The Straits Times* (ST, 26 January 2003a,b, 28 January 2003). These letters were also archived and posted on the Ministry of Foreign Affairs website (MFAS, 2003). Malaysia published a week-long series of advertisements on the issue, stating its position in the water talks (NST, 13 July 2003). This information was also compiled and made available on Malaysian government websites (NEAC, 2003). Furthermore, Malaysia made these copies available for sale to the public (NST, 20 July 2003). It is in these publications that the Singaporean and Malaysian media's framing of the water negotiations became more consistent and evident.

(b) Media Framing of the Singapore–Malaysia Water Relations

Both Singapore and Malaysia portrayed different perspectives of the water negotiations that became most evident in the third period. These perspectives were valid, but incomplete when viewed and read separately. An informed public needs to understand both sides of the situation. More often than not, however, the media seem to have reported what the two governments wanted their respective public to know about their respective policies.

Not surprisingly, both the Singaporean and the Malaysian governments portrayed themselves as the more reasonable partner in the bilateral talks. Each party claimed that the other was publishing inaccurate information. Malaysia claimed that Singapore published false information on the water issue (BH, 26 July 2003), and that its reports were misleading and did not accurately reflect the developments (*Sin Chew Daily* (SCD), 29 January 2003). Meanwhile, the Singaporean media reported how Malaysia ignored crucial facts (ST, 15 July 2003) and published false answers in its advertisement blitz (BH, 28 July 2003).

The divergent framing of the critical issues from Singapore and Malaysia is discussed in the next section.

The Singaporean Media on Water Relations: Singapore is Consistent and Reasonable

The Singaporean media portrayed Singapore as a consistent and reasonable negotiating partner in the water negotiations. It repeatedly framed Singapore as very accommodating, giving in to Malaysia's changing requests (BT, 16 October 2002; ST, 29 July 2002). It highlighted how Malaysia had been inconsistent and unreasonable in the negotiations: changing their nature, changing the agenda and changing the price of water.

– *Changing the Nature of Negotiations* The extension of the water contracts was first tied to a financial arrangement (BT, 6, 17 and 24 November 1998). It was then that Malaysia requested to drop the financial issues and instead adopted a package-deal approach to include a host of outstanding bilateral issues between the two countries (NST, 18 December 1998; ST, 24 July 2002). This approach linked the water issues to the early withdrawal of the Central Provident Fund for Malaysians, the relocation of the Malaysian Immigration checkpoint in Singapore and the use of Malaysian airspace by Singapore (BT, 18 December 1998). It was also Malaysia that added new issues into the agenda, such as the proposal for a new bridge to replace the Causeway (ST, 26 January 2003c; ST, 4 September 2002). In the end, it was also Malaysia that unilaterally dropped the package approach (ST, 3 July 2002; BT, 3 July 2002). Even then, Singapore was still willing to continue negotiations with Malaysia (ST, 24 July 2002 and 26 January 2003c).

– *Changing the Agenda of Water Negotiations* Focusing specifically on water, Malaysia often changed the "goalposts". The original talks were on the extension of the water agreements after the 1961 and 1962 contracts had expired (BT, 30 June and 6 November 1998; NST, 4 and 18 April 1998). Malaysia was the party that wanted to include the prevailing agreements on the table through a price revision for water, and for the price review to be retroactive (ST, 3 July and 24 July 2002; BT, 3 July 2002). Later on, it was also Malaysia that wanted to discuss only the price of current water and not of future water (BT, 9 and 12 October 2002; ST, 12 October 2002).

– *Changing the Water Price* Malaysia constantly changed its offer as to the price Singapore should pay for importing water (BT, 1 and 4 February, 8 July, 2002; ST, 8 July 2002). From an agreement of 45 sen per 1,000 gallons in 2000, Malaysia changed its asking price to 60 sen per 1,000 gallons in 2001 and then to RM 3 per 1,000 gallons (ST, 26 January 2003c). Together with the shift from a package approach to dealing with the water issue separately, the changing water price added to the protracted negotiations on water. It is important to mention that Singapore is subsidising the cost of treated water to Johor. Singapore is selling treated water at the rate of RM 0.5 per 1,000 gallons, although it is costing Singapore RM 2.4 per 1,000 gallons (http://www.mfa.gov.sg/internet/press/pedra/faq. html).

The Singapore media reported how Singapore had been reasonable and accommodating to Malaysia's changing requests (BT, 16 October 2002; ST, 29 July 2002). However, the negative publicity about the negotiations prompted Singapore to set the record straight by publishing official correspondence between the parties concerned (ST, 26 January 2003c). The speech of Foreign Minister Jayakumar and the additional information released through the letters further indicate how the media had framed the situation.

The Malaysian Media on Water Relations: Malaysia Wants a Fair Price

The Malaysian media portrayed Malaysia as the most reasonable partner, who only wanted a fair water agreement with Singapore (NST, 21 July 2003). Likewise, the publication of its booklet *"Water: The Singapore–Malaysia Dispute: The Facts"* (NEAC, 2003) also solidified this framing. This booklet highlighted how Malaysia had been a cooperative partner, willing to supply water to Singapore as long as a fair agreement was made.

– *Willingness to Supply Water to Singapore* Malaysia repeatedly expressed its sincerity in supplying water to Singapore even after the 1961 and 1962 agreements had expired (NST, 4 April and 5 August 1998; 2 September 2002; 7 February 2003). The governments of Pahang and Johor also expressed the same view (NST, 25 June 1998; ST, 21 January and 1 October 1999).

– *Desire for a Fair Agreement* Malaysia was just requesting a fair and reasonable arrangement with Singapore, and asking for a fair price (NST, 15, 19 and 21 July 2003, 29 January 2002). Instead, Singapore had been profiting from Malaysia due to the low cost of water supplied by Malaysia and the high price Singapore was charging for treated water subsequently supplied to Malaysia (NST, 13 March 2000, 10 January 2001, 22 January 2002). To make progress, Singapore should accept Malaysia's right for a price review (NST, 4 September 2002, 16 July 2003).

– *Reasonable Malaysia, Unbending Singapore* It was Singapore that was being unreasonable and refusing to accommodate Malaysian requirements which were fair and just (NST, 13 March 2000). Singapore repeatedly turned water into the pivotal issue in negotiations that forced Malaysia to delink water from the package approach (NST, 22, 23, 26 January 2002, 8 September 2002).

The Malaysian media reported how Malaysia was very reasonable toward Singapore, only asking for a fair price for its water. To counteract Malaysian aims, Singapore published official correspondence between the leaders of the two nations. This forced Malaysia to react and inform the Malaysian population on its views of the situation through an information campaign.

The Media, its Views and the Water Relations

When discussing the role of the media in the Singapore–Malaysia water relationships, it is essential to understand the historical and political context between the two countries at specific times.

Historically, water has always been considered a very important part of the Singapore–Malaysia bilateral agenda and, as such, water agreements have been signed in different years between 1927 and 1990. Later on, water became part of the so-called "package deals", where discussions included several other issues. In 1995, for example, water was discussed in connection to the Malayan Railway and the plans Malaysia had to invest in an electric train that would connect both countries. In 1998, water negotiations were linked to a financial recovery package that included the Malayan Railway land in Singapore, the use of Malaysian airspace by Singapore's aircrafts, the transfer of Malaysian shares no longer traded on CLOB International to the Kuala Lumpur Stock Exchange and the early release of the Central Providence Fund savings of Malaysian workers. In 2001, water was discussed together with the replacement of the Causeway bridge in favour of a railway tunnel that would connect both countries. Finally, in 2002, the bilateral talks were stalled, because it was proposed by Malaysia that water should be discussed separately from any package deal, and because of the discussions in regard to the right of Malaysia to revise the price of water well after the date stipulated in the water agreements signed by the two parties. During all this time, one would have expected the historical and political situations prevailing between the two countries to have influenced the views and opinions as well as the "sentiments", and even the "tone", of the media when covering the news, even if these were supposedly solely about water.

Overall, the main role of the media has been to publicise and inform the public on the water negotiations, mainly from the viewpoints of the respective national interests, which is to be expected on bilateral issues. This publicity has been effective to the point that other groups, both private and public, that were not party to the official negotiation process, expressed their own thoughts and opinions on what their govern-ments should do. For example, there was an increase in the number of interest groups, such as the UMNO youth group, the PAP (People Action Party) (Yap *et al.*, 2010) youth group, the PAP Opposition party, NGO groups, the Malaysian military and research institutions, that expressed their opinions on the issue. The increased media coverage and the larger number of interested stakeholders even prompted the Singa-porean and Malaysian Foreign Ministries to explain the details of the protracted talks to their citizens. In fact, the complexity created by the emergence of so many new players impelled the officials to ask the public to stop interfering in this matter (SCD, 20 January 2003).

As the discussions on water between Singapore and Malaysia evolved over time, the views of the media and their coverage on the topic of water also changed, shaping the public perception on this issue in both countries. Initially, media coverage in both countries played both an informative and a constructive role, with positive images of cooperation and consistent framing in the news. This was a period that was dominated by news articles. Later on, the media became a vehicle of communication for interested groups other than officials, publishing news articles, opinions, editorials, forums and letters. Media coverage was consistent, with a slight amount of news framing, which gradually became more negative in both the countries until the framing of the issues differed. Finally, the 2002–2004 period witnessed the most intense coverage on the water discussions, when the media became a vehicle of communication for officials in both the countries (with news articles, opinions, editorials, forums, letters and pamph-lets). The frame of the news was very different in the two countries, with predominantly negative news on both sides of the Causeway. During this last period, the perspectives presented by the media of both countries were incomplete, when read separately, with the net result that the readers were getting only a partial view of the situation.

By and large, the local interest in the two countries grew with the proliferation of opinions and editorial articles. When bilateral negotiations became a domestic political issue, the leaders used the media to clarify and explain the status of negotiations to their respective public, trying to make their views clear for their own citizens, but also sending "unofficial" messages to interested parties in the other country. Clearly, with the growing interest in the water negotiations, the media also served as a platform for the countries not only to inform their citizens but also to educate them on the progress of the negotiations.

As mentioned earlier, mass media coverage of political issues is normally, and necessarily, selective. The media normally depends on frames to give coherence to relatively brief treatments of complex issues through certain views, in which they select to highlight specific items and ignore others, often looking for consensus or disagree-ment on certain matters. Nevertheless, as important as the media is to shape public opinion, and as consistent as the messages of the media can be with their respective establishments, citizens normally make use of their own perspectives, listening to some information and disregarding some other, depending on whether they agree or not with the views presented. It so happens that citizens often interpret media content to rein-force their own views.

The view on the role of the media can be more comprehensive if considered within its multifaceted relationship with the State and the public, with the three actors being equally important in the equation. In the water relationship between Singapore and

Malaysia, the media has been a dynamic actor—a role which has evolved with time during the course of the bilateral negotiations—but also has taken on the attitude of the citizens in both countries. While the media of each country has portrayed the water negotiations in different lights, playing important roles in terms of reporting on policies and politics, the readers have also played an important role mainly in terms of supporting their own countries, often willingly accepting the viewpoint of the establishment, considering it to be acting according to their national interests. The role of the media would be partial only if considered in isolation, without acknowledging the role of the readers, since, frequently, both of them voice the ideas, ideals and concerns of the other.

Further Thoughts

Relationships between Singapore and Malaysia have "undergone a sea change" during the last decade or so (Chang *et al.*, 2005, p. 1), with bilateral relations improving significantly with the change of leadership in Malaysia in 2003. Stronger bilateral ties have been characterised by greater contact and cooperation between the leaders, officials and businesses. Synergies are developing and include, but are not limited to: economic cooperation, trade and investment; increasing private sector participation in strategic investments, corporate purchases and joint business ventures; cooperation on security matters; movement of technical experts across borders; promotion of tourism and sport-related activities; exchange of students; and improved relationships between civil society groups in both countries (Saw and Kesavapany, 2006; Sidhu, 2006). With the objective to promote better understanding and bilateral ties among the citizens of both countries, circulation of newspapers on "both sides of the Causeway" has been re-initiated after some 30-year ban of each others' newspapers (Saw and Kesavapany, 2006, p. 17).

Bilateral negotiations have also resumed with the aim to solve the several outstanding issues, considering all the various alternatives. In 2004, it was agreed, by Singapore's former Prime Minister Goh and then Malaysian Prime Minister Abdullah Badawi, that future discussions between the two countries should be based on the consideration of mutual benefits on any proposal that would be discussed. It was also agreed that the outstanding issues yet to be resolved should not delay cooperation in other areas (Saw and Kesavapany, 2006). With the political environment more positive between the two countries, relations are in a new phase where both the countries want to resolve bilateral differences. This has already resulted, for example, in the amicable solution of the land reclamation dispute. In April 2005, Malaysia withdrew the case against Singapore from the International Tribunal of the Law of the Seas. Singapore agreed to make adjustments to its reclamation works, and compensated Malaysian fishermen for losses due to the works. Both countries signed an agreement, on 26 April 2005, that Johor Straits form a "shared water body" (Sidhu, 2006, p. 88). Malaysia also discontinued the project on the bridge to replace the Causeway on 12 April 2006. Finally, the dispute on the sovereignty of Pedra Branca was also solved when the International Court of Justice ruled in favour of Singapore.

In terms of the media, it should be noted that Singapore and Malaysia resumed talks on outstanding bilateral issues in 2005, and decided not to divulge the details of the negotiations through the media (MFAS, 2005). Both countries have agreed that details of the negotiations on water should not be discussed with the media. It has been recognised that it would not be helpful to publicise the details of the negotiations to avoid heightening expectations, as happened on earlier occasions, as well as to avoid media frenzy on whatever issues had been discussed, as had been the case in the past (and would be the case once more in the media reports on the "scenic bridge" case presented in Saw and Kesavapany, 2006, p. 6–7).

Former Singapore Prime Minister Goh is quoted to have said that "due to the sensitive nature of issues both sides have agreed to keep discussions in private instead of negotiating through the press—as it has been the case in the past" (Agence France-Presse, 17 October 2004, in Sidhu, 2006, p. 87). The low-key and private nature of the discussions that have been held are considered as clear signals of the willingness of both countries to solve bilateral problems between them, as well as to avoid the media capitalising on these issues, as it did in the past (Sidhu, 2006). A private setting or "quiet diplomacy" has clearly been accepted by both the countries as the best alternative to achieve progress (Lee, 2010).

The ties between both the countries are deep-rooted and are based not solely on water but on a multiplicity of others factors. As noted by Mr Tan Gee Paw, current Chairman of the Public Utilities Board of Singapore, "There is much that both countries can gain by working together. Our common interests far exceed our bilateral differences" (BBC World Debate on Water, Singapore, 30 June 2010).

Acknowledgements

This chapter is part of a broader study on the historical perspectives of urban water management in Singapore. It is funded by the Public Utilities Board and conducted within the framework of the Lee Kuan Yew School of Public Policy, NUS, Singapore, and Third World Centre for Water Management, Mexico. The authors would like to thank Prof. Cherian George, Associate Professor at the Wee Kim Wee School of Communication and Information (http://www.wkwsci.ntu.edu.sg), Nanyang Technological University and Adjunct Senior Research Fellow of the Institute of Policy Studies (http://www.spp.nus.edu.sg/ips/home.aspx) at the Lee Kuan Yew School of Public Policy, and Ms Ching Leong, PhD Student at the Lee Kuan Yew School of Public Policy, for their insightful comments on the role of media on policy making in Singapore.

Notes

* The older spelling "Johore" is used in some references.
1 NEWater is recycled wastewater which has become a fundamental resource to ensure secure and sustainable use of water in Singapore (see Tortajada, 2006; Soon *et al.*, 2009; Ching, 2010; Luan, 2010)

Newspaper References

BH (*Berita Harian*) (26 July 2003) Water advertisement: Singapore continues to broadcast the wrong facts. Translated from Malay into English.
BH (28 July 2003) MTEN broadcasts in response to Singapore's mis-reports in AWSJ. Translated from Malay into English.
BT (*The Business Times*) (6 June 1997) Framework for Wider Cooperation.
BT (10 April 1998) Anwar clarifies changes on Bumiputra ownership rules, Yang Razali Kassim.

BT (30 June 1998) S'pore, KL have not reached accord on water.

BT (4 August 1998) Suspend fresh ties with S'pore, urges Youth chief. Eddie Toh.

BT (21 August 1998) PUB embarking on various projects to ensure long-term water supply. Ronnie Lim.

BT (6 November 1998) M'sia seeks Singapore's help to raise funds. Ruth Wong.

BT (17 November 1998) 'Convergence of views' on S'pore-M'sia Cooperation. Yang Razali Kassim.

BT (24 November 1998) KL has sent S'pore draft loan agreement: PM Goh. Anna Teo.

BT (18 December 1998) KL passes up US$4B S'pore loan, ties water to other issues KL seen sorting out Clob transfer soon.

BT (11 June 1999) Looking Beyond M'sia for water.

BT (15 June 2000) S'pore Poh Lianuying into Indon water project. Shoeb Kagda.

BT (19 August 2000) Johor to build new RM 700m water treatment plant. Eddie Toh.

BT (15 February 2001) New initiatives to boost economic ties with Riau. Shoeb Kagda.

BT (24 April 2001) Johor to have better water management. Eddie Toh.

BT (5 September 2001) SM Lee, Dr M strike in-principle accord on bilateral issues.

BT (1 February 2002) Malaysia will insist on higher rate for water to S'pore: Dr M; A review of the in-principle pact reached last September now looks likely.

BT (4 February 2002) HK's high raw water price includes infrastructure costs; Johor has not borne any such costs for S'pore supply: MFA. Audrey Tan.

BT (3 July 2002) Malaysia to review retroactively price of water; KL takes water, new bridge talks out of package of unresolved issues. Eddie Toh.

BT (8 July 2002) M'sia seeks up o RM3 per thousand gallons of water; It also wants to incorporate new formula after 2011. Eddie Toh.

BT (12 July 2002) NEWater gets panel's clearance; 2-year study clears flow of reclaimed water into reservoirs. Chuang Peck Ming.

BT (17 July 2002) NEWater purer than PUB's. Chuang Peck Ming.

BT (26 July 2002) Good enough to quench the thirst.

BT (9 October 2002) Water price review must be part of package: PM; Mr Goh: S'pore willing to meet M'sia's wish if it's within package. Eddie Toh.

BT (12 October 2002) KL wants to discontinue package approach on outstanding issues; Dr M wants to backdate revised price of water. Eugene Low and Eddie Toh.

BT (16 October 2002) KL has no legal right to seek water price review: S'pore. Chuang Peck Ming.

BT (25 October 2002) M'sia mulls new laws to dilute water pacts; Minister says law could allow Johor to determine amount to supply to S'pore. Addy Toh.

NST (New Straits Times) (4 April 1998) KL and Singapore set to resolve water supply issues.

NST (18 April 1998) Understanding over water supply to Singapore after 2061 reached. Ashraf Abdullah.

NST (25 June 1998) Pahang to consult NWC on sale of water to Singapore. Azran Aziz and Sharif Haron.

NST (4 August 1998) Statements do not reflect Republic's true stance. Sharif Haron.

NST (5 August 1998) Dr M: We will not cut water supply to Singapore.

NST (18 December 1998) Republic's RM 15.2b loan for water not needed. Ashraf Abdullah.

NST (23 February 1999) It's up to Singapore to offer solutions to outstanding problems. Letters section by C.C. Chin.

NST (7 June 1999) Ghani: Our need for water to be given priority. Charlotte Venudran.

NST (8 June 1999) Local water needs the priority. NST (17 June 1999) Johor also benefits from water agreement. Samuel Tan.

NST (17 June 1999) Johor also benefits from water agreement. Samuel Tan.

NST (13 March 2000) Syed Hamid denies that UMNO elections holding up talks with Singapore.

NST (10 January 2001) Present water treaty not in favour of Malaysia.

NST (17 January 2001) We should not depend on others to supply our basic water needs. Mohamed Maharis.

NST (22 January 2002) Discussion with Singapore stalled due to water issue. Ramlan Said.

NST (23 January 2002) Settle water issue quickly.

NST (26 January 2002) DPM: Priority for resolving water issue.

NST (29 January 2002) Malaysia just as eager to solve bilateral issues. Cheah Chor Sooi.

NST (6 August 2002) Singapore plans to let first of two water deals lapse in 2011.

NST (7 August 2002) More favour than trade. Firdaus Abdullah and M.K. Megan.

NST (2 September 2002) Malaysia still willing to pump millions of gallons of scarce water to Singapore. Manan Osman.

NST (4 September 2002) Singapore agrees to discuss price. Manan Osman.

NST (8 September 2002) MB: Singapore's stand on water issue a stumbling block to ties. Chong Chee Seong.

NST (14 October 2002) Malaysia to seek legal recourse if no headway made on water price issue. Carol Murugiah.

NST (30 November 2002) Water issue: Malaysia to stop talks with Singapore. Firdaus Abdullah.

NST (28 January 2003) Singapore action criticized (HL). Rahman Said.

NST (7 February 2003) Syed Hamid: We're unhappy with pricing but we have honoured water pacts. Carolyn Hong and Firdaus Abdullah.

NST (13 July 2003) Singapore RM 662m profit. Ramlan Said.

NST (15 July 2003) Only a few cents for water. Ramlan Said.

NST (16 July 2003) Malaysia has right to review. Ramlan Said and Freeda Cruez.

NST (19 July 2003) NEAC: Asking for fair price is not bullying.

NST (20 July 2003) Booklet will tell the truth. Saiful Azhar Abdullah

NST (21 July 2003) Is fair price too much to ask?

SCD (*Sin Chew Daily*) (20 January 2003) Topics between Singapore and Malaysia awaiting governments to be resolved.

SCD (29 January 2003) The water negotiations demands are misleading: not reflecting true stories.

ST (*The Straits Times*) (16 January 1999) Water deal with Jakarta is possible, says Philip Yeo. Susan Sim.

ST (21 January 1999) Cheaper to buy water from Johor, MB tells Singapore.

ST (3 March 1999) Malaysia can choose not to buy treated water.

ST (1 October 1999) Pahang wants to sell water to boost coffers.

ST (2 July 2000a) Massive water project is floated. Yeoh En-Lai and Liang Hwee Ting.

ST (2 July 2000b) Riau in Sumatra keen to fill S'pore's water needs. Yeoh E n-L ai.

ST (2 December 2000) End reliance on neighbors for supply.

ST (15 March 2001) Soon: cheaper to desalinate seawater than import it. Sharmilpal Kuar.

ST (22 March 2001) 30m gallons a day to drink, from the sea. Sharmilpal Kuar.

ST (5 September 2001) Thorny issues that go back many years. Irene Ng and Brendan Pereira.

ST (7 September 2001) WP welcomes new water deal.

ST (3 July 2002) KL seeking to settle water pricing separately. Chua Lee Hoong.

ST (8 July 2002) Malaysia reveals asking price for water; 60 sen per 1000 gallons till 2007, after which it will be raised to RM 3, says Foreign Minister. S'pore pays 3 sen now. Reme Ahmad.

ST (12 July 2002) Experts find reclaimed water safe to drink; International panel gives Singapore's NEWater thumbs up after 2-year study; nod for blending it with reservoir water. Dominic Nathan.

ST (13 July 2002) NEWater can replace Johor supply; DPM Lee says water bought elsewhere must be competitive with reclaimed water, which is a 'serious alternative' Ginnie Teo.

ST (24 July 2002) Bilateral issues resolved only as a package; Foreign Minister S. Jayakumar updated MPs on the bilateral issues discussed when he met his counterpart, Malaysian Foreign Minister Syed Hamid Albar, in Kuala Lumpur on July 1 and 2.

ST (29 July 2002) Water: A toast to more comfortable bilateral dealings. Asad Latif.

ST (11 August 2002) Johor leads pokes fun at NEWater; He warns Malaysians that they risk drinking water recycled from washrooms when in Singapore.

ST (4 September 2002) KL-S'pore talks hit snag as Malaysia changes tack; Malaysia now wants water as the issue of the Customs checkpoint to be dealt with separately, instead of as a package. Tan Tarn How.

ST (3 October 2002) Sell sewage to S'pore instead, says MP.

ST (12 October 2002) KL no longer wants to settle issues as package; Mahathir states this in a letter to PM Goh; he also wants to backdate any price hike of water to 1986 and 1987. Brendan Pereira Lydia Lim.

ST (25 October 2002) KL mulls over new law to scrap water accords; Act could render supply of water to other countries subject to Malaysia's domestic needs; move to be considered if talks fail. Reme Ahmad.

ST (20 November 2002) KL insists it will discuss only water-price review. Brendan Pereira.

ST (21 November 2002) Singapore restates stand on water talks; both the current price and future water supply should be discussed, Republic says in response to KL's latest remarks.

ST (1 December 2002) S'pore 'waiting for KL's clarification on water talks'.

ST (3 December 2002) KL hints at other options to settle water issue; It refuses to confirm that it is seeking arbitration over the matter. Rene Ahmad.

ST (26 January 2003a) Dear Kuan Yew.

ST (26 January 2003b) Letters tell the true story.

ST (26 January 2003c) What is at stake: our very existence as a nation.

ST (28 January 2003) Dear Mahathir.

ST (15 July 2003) KL's water ad ignores crucial facts, says Singapore; Foreign Minister says it's a rehash of an old arguments and is puzzled by the timing of the current campaign against the Republic. Rebecca Lee.

ST (2 August 2003) Water supply deal will remain: Mahathir; Federal control over resources will not affect S'pore supplies, he says, but adds that time for talks is over.

The Star (18 August 2002) Minister: NEWater won't affect our stand.

References

Andina-Díaz, A., 2007. Reinforcement *vs* change: the political influence of the media. *Public Choice*, 131, 65–81.

Ang, P.H., 2002. The media and the flow of information. *In*: Derek Da Cunga, ed. *Singapore in the New Millennium. Challenges facing the city state*. ISEAS, Singapore, 243–268.

Ang, P.H., 2007. *Singapore Media. Journalism.org*. Available from: http://journalism.sg/wp-content/uploads/2007/09/ang-peng-hwa-2007-singapore-media.pdf [Accessed 10 January 2011].

Agreement as to Certain Water Rights in Johore between the Sultan of Johore and the Municipal Commissioners of the Town of Singapore signed in Johore in 5 December 1927. Available from: http://www.mfa.gov.sg/kl/doc.html [Accessed 15 March 2011].

Agreement between the Government of the State of Johor and the Public Utilities Board of the Republic of Singapore, signed in Johore on 24 November 1990. Available from: http://www.mfa.gov.sg/kl/doc.html [Accessed 15 March 2011].

Chang, C.Y., Ng, B.Y., and Singh, P., 2005. Roundtable on Singapore– Malaysia Relations: mending fences and making good neighbours. *Trends in Southeast Asia Series*, 16.

Ching, L., 2010. Eliminating Yuck: a simple exposition of media and social change in water reuse policies. *Internatinal Journal of Water Resources Development*, 26 (1), 111–124.

Department of Statistics, 2010. *Year Book of Statistics, Malaysia, 2009*. Kuala Lumpur: Government of Malaysia.

Department of Statistics, Ministry of Trade & Industry, 2010. *Yearbook of Statistics*. Singapore: Government of Singapore.

Dhillon, K.S., 2009. *Malaysian foreign policy in the Mahathir Era 1981–2003, Dilemmas of Development*. Singapore: NUS Press.

George, C., 2007. Singapore's emerging informal public sphere: a new Singapore. In: Tan Tarn How, ed. *Singapore Perspectives 2007*. Singapore: Institute of Policy Studies and World Scientific.

Ghesquière, H.C., 2007. *Singapore's Success: Engineering Economic Growth*. Singapore: Thomson Learning .

Government of Singapore, 1974. *Newspaper and Printing Presses Act*. Available from: http://statutes.agc.gov.sg/non_version/ cgi-bin/cgi_retrieve.pl?actno=REVED-206&doctitle=NEWS PAPER%20AND%20PRINTING%20PRESSES%20ACT%0A &date=latest&method=part [Accessed on 20 January 2011].

Guarantee Agreement between the Government of Malaysia and the Government of the Republic of Singapore signed in Johore in 24 November 1990. Available from: http://www.mfa.gov.sg/kl/doc.html [Accessed 15 March 2011].

Johore River Water Agreement between the Johore State Government and City Council of Singapore signed in Johore in 29 September 1962. Available from: http://www.mfa.gov.sg/kl/doc.html [Accessed 15 March 2011].

Kellner, D., 2000. Habermas, the public sphere and democracy: a critical intervention. *In*: L.E. Hahn, ed. *Perspectives on Habermas*. Open Court: Chicago and La Salle, Illinois, 259–538.

Kenyon, A., 2007. Transforming Media Market: The Cases of Malaysia and Singapore. *Australian Journal of Emerging Technologies and Society*, 5 (2), 2007, 103–118. Available from: http://www.swin.edu.au/hosting/ijets/journal/V5N2/pdf/ Article3-KENYON.pdf [Accessed on 23 March 2010].

Kim, W., 2001. Media and Democracy in Malaysia. Media and Democracy in Asia. *The Public*, 8 (2), 67–88. Available from: http://www.nstp.com.my/Corporate/nstp/products/product Sub. htm [Accessed on 20 January 2011]

Kog, Y.C., 2001. *Natural resource management and environmental security in Southeast Asia: case study of clean water supplies in Singapore*. Singapore: Institute of Defence and Strategic Studies.

Kog, Y.C., Jau, I.L.F., and Ruey, J.L.S., 2002. Beyond Vulnerability? Water in Singapore–Malaysia Relations, *IDSS Monograph No. 3*. Singapore: Institute of Defence and Strategic Studies.

Lee, P.O., 2003. The water issue between Singapore and Malaysia, no solution in sight? *Economic and Finance*, 1, Institute of Southeast Asian Studies, Singapore.

Lee, P.O., 2005. Water management issues in Singapore. *Paper presented at Water in Mainland Southeast Asia*, 29 November–2 December 2005. Siem Reap, Cambodia, Conference organised by the International Institute for Asian Studies (IIAS), Netherlands, and the Centre for Khmer Studies (CKS), Cambodia.

Lee, P.O., 2010. The four taps: water self-sufficiency in Singapore. In: T. Chong, ed. *Management of success: Singapore revisited*. Singapore: Institute of Southeast Asian Studies, 417–439.

Long, J., 2001. Desecuritizing the water issue in Singapore–Malaysia relations. *Contemporary Southeast Asia*, 23 (3), 504–532.

Luan, I.O.B., 2010. Singapore Water Management Policies and Practices. *International Journal of Water Resources Development*, 26 (1), 65–80.

MFAS (Ministry of Foreign Affairs of Singapore), 2003. *Singapore Government. Statement by Minister for Foreign Affairs*, Prof. S. Jayakumar, In: Parliament, 25th January 2003. Available from: http://www.mfa.gov.sg/internet/press/water/speech.html#an nex [Accessed 15 March 2010].

MFAS (Ministry of Foreign Affairs of Singapore), 2005. *Joint Press Release on the meeting between Malaysia and Singapore on the outstanding bilateral issues.* Available from: http://app. mfa.gov.sg/2006/press/view_press.asp?post_id= 1258 [Accessed 29 June 2010].

MICA (Ministry of Information, Communications and the Arts) 2003. Ministerial Statement by Prof. S. Jayakumar, Singapore Minister for Foreign Affairs in the Singapore Parliament on 25 January 2003. In: *Water Talks? If Only It Could.* Ministry of Information, Communications and the Arts, Singapore. Annex A, 67–80. Available from: http://www.mfa.gov.sg/ internet/ press/water/speech.html#annex [Accessed 20 July 2010].

NEAC (National Economic Action Council), 2003. *Water: The Singapore–Malaysia Dispute: The Facts. National Economic Action Council, Kuala Lumpur.* Available from: http:// thestar. com.my/archives/2003/7/21/nation/waterbooklet3.pdf [Accessed 15 March 2010].

Saw, S.H. and Kesavapany, K., 2006. *Singapore–Malaysia relations under Abdullah Badawi.* Singapore: Institute of Southeast Asian Studies.

Shiraishi, T. (ed.) 2009. *Across the causeway. A multidimensional study of Malaysia–Singapore relations.* Singapore: Institute of Southeast Asian Studies.

Shriver, R., 2003. *Malaysian media: ownership control and political content.* Available from: http://www.rickshriver.net/ Documents/Malaysian%20Media%20Paper%20-%20CARFAX2. pdf [Accessed 3 April 2010].

Sidhu, J.S., 2006. Malaysia–Singapore relations since 1998: a troubled past—whither a brighter future? In: R. Harun, ed. *Malaysia's foreign relations. issues and challenges.* Kuala Lumpur: University Malaya Press, 75–92.

SPH (Singapore Press Holdings), 2010. *Singapore press holdings annual report 2009. Growing with the times.* Available from: http://www.sph.com.sg/annual_report.shtml [Accessed 15 January 2011].

Soon, T.Y., Lee, T.J., and Tan, T., 2009. *Clean, Green and Blue.* Singapore: Institute of Southeast Asian Studies and Ministry of Water and Environment Resources.

Tan, T.H., 2010. Singapore's print media policy. A national success? *In*: T. Chong, ed. *Management of success. Singapore revisited.* ISEAS: Singapore, 242–256.

Tebrau and Soudai Rivers Agreement between the Government of the State of Johore and the City Council of the State of Singapore signed on 1 September 1961. Available from: http:// www.mfa.gov.sg/kl/doc.html [Accessed on 15 March 2011].

Tortajada, C., 2006. Water Management in Singapore. *International Journal of Water Resources Development.* 22 (2), 227–240.

World Bank, 2010. *World Development Indicators Database 2008.* Available from: http://siter-esources.worldbank.org/ DATASTATISTICS/Resources/GNIPC.pdf [Accessed 20 July 2010)].

Yap, S., Lim, R., and Kam, L.W., 2010. *Men in White: The Untold Story of Singapore's Ruling Political Party.* Singapore Press Holdings Limited and Marshall Cavendish International (Asia) Private Limited: Singapore.

Appendix A
Singapore–Malaysia Water Agreements

The water relationship between Singapore and Malaysia dates as far back as 1927, decades before both countries gained independence. Even though the water relations have not always been perfect, the four water agreements illustrate a history of consistent cooperation between the two countries in the issue of water. These agreements, of which only the 1927 one is no longer in force, allow Singapore to source water from Malaysia, and for Malaysia to buy back treated water from Singapore. This relationship has secured water supply for Singapore and brought water infrastructure development to Malaysia through Singapore investments in Johor.

The following table summarises the water agreements between Malaysia and Singapore.

Table 1 Water Agreements between Singapore and Malaysia.

Water Agreements	Parties to the agreement	Water agreements details
1927	The Sultan of Johore and the Municipal Commissioners of the Town of Singapore	The agreement allows Singapore to rent 2100 acres of land in Gunong Pulai at 30 cents per acre per year, and "take, impound and use all the water which from time to time may be or be brought or stored in upon or under the said land" at no cost to the municipality. Singapore also had the right to lay and maintain the necessary waterworks to transfer the water. The Government of Johore could request the supply of 800 000 gallons of water per day, if necessary, at 25 cents per 1000 gallons.
1961	The Johore State Government and the City Council of the State of Singapore	Under this agreement, the Government of Johore reserved the lands, hereditaments and premises situated at Gunong Pulai, Sungei Tebrau and Sungei Scudai in the State of Johore for the use by the City Council. The City Council shall pay to the Government an annual rent of $5 per acre. The Government of Johore shall not for a period of 50 years alienate or do any act of deed affecting the said land or any part thereof during such term. The Government of Johore agreed to give the City Council the full and exclusive right and liberty to enter upon and occupy the land, and take, impound and use all water from the Tebrau and Scudai rivers, as well as to construct the necessary water works, reservoirs, dams, pipelines, aqueducts, etc. The City Council would supply to the Johore Government, if and when requested by the Government, a daily amount of water not exceeding at any time 12% of the total quantity of water supplied to Singapore over the Causeway, and in no case less than 4 million gallons. The quality of the water would have to be of accepted standard and fit for human consumption. The City Council would pay to the Government, 3 cents for every 1000 gallons of water drawn from the State of Johore and delivered to Singapore, and the Government of Johore would pay to the City Council 50 cents for every 1000 gallons of pure water. When the City Council had to provide the Johore Government with raw water, it would pay 25 cents for every 1000 gallons of the water supplied. These clauses are subject to review after the expiry of 25 years time. Prices can be revised in the light of any change in the purchasing power of money, cost of labour and power and material for the purpose of supplying the water. In the event of any dispute or differences arising under the provisions of this clause the same shall be referred to arbitration as provided in the agreement.
1962	The Johore State Government and City Council of Singapore	The Government of Johore agreed to demise unto the City Council all and singular specific lands in the State of Johore for a period of 99 years. The Government granted the City Council "the full and exclusive right and liberty to draw off, take, impound and use the water from the Johore River up to a maximum of 250 million gallons per day". The City Council would supply to the Government a daily amount of water drawn off from the Johore River not exceeding at any time 2% of the total quantity of water supplied to Singapore, the quality of which would always be of acceptable standard and fit for human consumption. The City Council shall pay to the Government 3 cents for every 1000 gallons of water drawn from the Johore River and delivered to Singapore. The Government would pay to the City Council 50 cents for every 1000 gallons of pure water supplied. Should it be necessary for the City Council to supply raw water to the Government, this would pay to the City Council 10 cents for every 1000 gallons of the raw water. The price can be revised after the expiry of the agreement in 25 years' time, and in line with the rise or fall in the purchasing power of money, cost of labour, and power and materials for the purpose of supplying the water. In the event of any dispute or differences arising under the provisions of this clause the same shall be referred to arbitration as provided by the agreement.
1990	The Government of the State of Johor* and the Public Utilities Board of the Republic of Singapore	The Johor Government agreed to sell treated water generated from the Linggiu Dam to PUB in excess of the 250 million gallons per day of water under the 1962 Johor River Water Agreement considering that PUB agreed "to build at its own cost and expense the Linggiu Dam and other ancillary permanent works in connection therewith and thereafter to run, operate and maintain at its own cost and expense the dam, reservoir and

(Continued)

35

Table A1 (*Continued*)

Water Agreements	Parties to the agreement	Water agreements details
		ancillary permanent works". The Agreement shall expire upon the expiry of the 1962 Johore River Water Agreement. However, it can be extended beyond the original terms should the parties agreed to it. It was agreed that PUB shall purchase treated water at the price of either the weighted average of Johor's water tariffs plus a premium which is 50 percent of the surplus from the sale of this additional water by PUB to its consumers after deducting Johor's water price and PUB's cost of distribution and administration of this additional water, or 115 percent of the weighted average of Johor's water tariffs, whichever is higher". The quality of the treated water supplied to PUB under this Agreement shall conform with the prevailing World Health Organization's guidelines for drinking water. In terms of land, the Johor Government agreed that the State land to be used for the catchment area and the reservoir of approximately 21,600 hectares shall be leased for the remaining period of the 1962 Johore River Water Agreement. The premium for the land shall be calculated at the rate of M\$18,000 per hectare and an annual rent at the rate of M\$30 for every 1000 square feet of the said land. The annual rent will be subject to any revision imposed by the State Authority under the provisions of the National Land Code of Malaysia (Act No. 56/65). PUB agrees to pay M\$320 million as compensation for the permanent loss to the use of the land, the loss of revenue from logging activities in the form of premium, royalty and cess payment and the one-time up front payment for the leasing of the said land, inclusive of rentals for the remaining tenure of the 1962 Johore River Water Agreement. Any dispute or difference between the parties which cannot be resolved amicably by discussions between the parties shall be settled by arbitration in accordance with the Rules of the Regional Centre of Arbitration at Kuala Lumpur at that time.

* Spelling as in the Agreement.
Sources: see the several water agreements and the Constitutions of Singapore and Malaysia for further information: The Agreement as to Certain Water Rights in Johore between the Sultan of Johore and the Municipal Commissioners of the Town of Singapore signed in Johore on 5 December 1927; The Johore River Water Agreement between the Johore State Government and City Council of Singapore signed in Johore on 29 September 1962; The Tebrau and Soudai Rivers Agreement between the Government of the State of Johore and the City Council of the State of Singapore signed on 1 September 1961; Agreement between the Government of the State of Johor and the Public Utilities Board of the Republic of Singapore, signed in Johore on 24 November 1990; Guarantee Agreement between the Government of Malaysia and the Government of the Republic of Singapore signed in Johore on 24 November 1990.

Appendix B
Chronology of Key Developments

17 December 1998	PM Goh agreed with PM Mahathir's proposal to resolve outstanding bilateral issues, including long-term supply of water to Singapore, together as a package.
March–May 1999	Officials from both sides met three times, but made little progress.
15 August 2000	At a four-eye meeting in Putrajaya, SM Lee and PM Mahathir reached an agreement on a list of items including the price of 45 sen per 1000 gallons for current and future water. This was the first time the issue of current water was discussed as part of the package. Singapore also agreed to discuss Malaysia's proposal to build a new bridge to replace the Causeway as part of the package.
24 August 2000	SM Lee wrote to then Malaysian Finance Minister Tun Daim Zainuddin confirming the list of items which he and PM Mahathir had agreed to.
21 February 2001	PM Mahathir replied to SM Lee that "Johore believes that a fair price would be 60 cents (*sic*) per mgd (*sic*) of raw water" and that this "should be reviewed every five years" (He meant 60 sens per thousand gallons).
23 April 2001	SM Lee noted in his reply to PM Mahathir that there were two main variations from their oral understanding reached in August 2000. These were PM Mahathir's proposal of 60 sen for raw water and the mix of raw and treated water to be supplied.
4 September 2001	SM Lee met with PM Mahathir for a second time in Putrajaya. At a joint press conference, they announced that they had agreed on the basic skeleton of an agreement. SM Lee explained that Singapore had offered to pay 45 sen for current raw water in return for assured raw water supply from Malaysia beyond 2061.

(*Continued*)

APPENDIX B (*Continued*)

8 September 2001	SM Lee wrote to PM Mahathir to follow-up on their September 4, 2001 discussion on Malaysia's proposal for a bridge to replace the Causeway.
21 September 2001	SM Lee wrote again to PM Mahathir concerning other issues in the package. Singapore affirmed its proposals to revise the price of current water from 3 sen to 45 sen per 1000 gallons, in return for Malaysia agreeing to supply water, at 60 sen, beyond the expiry of the existing agreements, in 2011 and 2061. The 60 sen price would be reviewed every five years for inflation.
18 October 2001	PM Mahathir now said Johor wanted 60 sen for water sold to Singapore. He also suggested that Singapore compensate Malaysia with more land parcels, should the KTM rail service end in Johore Baru.
10 December 2001	SM Lee replied to PM Mahathir to clarify Singapore's proposal on the bilateral issues and to seek clarification on the additional railway lands referred to by PM Mahathir. He expressed the hope that PM Mahathir would consider the long-term significance and value of retaining the railway link between Malaysia and Singapore. He requested PM Mahathir to set out Malaysia's position on the package of issues so as to establish a clear framework for officials to work on.
5 February 2002	Prompted by repeated Malaysia comments to the media that existing Water Agreements were unfair, Singapore conveyed a Third Party Note to register its deep concern over those remarks.
4 March 2002	PM Mahathir conveyed yet another new pricing proposal for water – this time a three-stage proposal. The asking price was now 60 sen for water from 2002 to 2007, RM3 from 2007 to 2011, and RM3 adjusted for inflation every year after 2011. As for the treated water Johor now buys from Singapore, Malaysia proposed that the price be raised simply from the current 50 sen to RM1, with no price review mechanisms.
11 March 2002	SM Lee wrote to Dr Mahathir noting that the latest proposals had changed completely from those agreed upon early. He would therefore have to study the implications of Malaysia's new offers before responding.
11 April 2002	PM Goh wrote to Dr Mahathir, conveying Singapore's response to the latests proposals. He said that for the sake of good long-term relations, Singapore would supplement the existing Water Agreements by producing its own NEWater. Since Malaysia did not accept Singapore's earlier offer of 45 sen for current water and 60 sen for future water, Singapore proposed to peg the price of future water to an agreed percentage of the cost of alternative source of water, which was NEWater. He also suggested that PM Mahathir's letter of March 4, 2002 and PM Goh's reply of April 11, 2002 form the basis for further discussions between the respective Foreign Ministers and officials.
1–2 July 2002	The two Foreign Ministers and their officials met on Putrajaya. Malaysia invoked the Hongkong (spelling in the printed text) model, in which Hongkong pays China RM8 per thousand gallons for its water. Singapore said it was willing to negotiate a price review provided this is done as part of a package even (t)hough it believes Malaysia's right to review expired in 1986 and 1987. It also pointed out that unlike Singapore, Hongkong does not have to bear the infrastructure and maintenance costs of drawing water.
2–3 September 2002	The Foreign Ministers met a second time, in Singapore. This time, Malaysia proposed a formula that resulted in a price of RM6.25 per 1000 gallons for current raw water. Malaysia also proposed that discussions on future water take place only in 2059.
8 October 2002	PM Mahathir told PM Goh while both were in Putrajaya that Malaysia wanted to "decouple the water issue" from the other items in the package. PM Goh responded that if the water issue was taken out of the package, then Singapore would have less leeway to make concessions on other issues.
10 October 2002	PM Goh received a letter from PM Mahathir dated October 7, 2002, in which Malaysia declared that it had decided to discontinue the package approach. Dr Mahathir did not mention that he had written this letter when he spoke to Mr Goh on October 8, 2002.
14 October 2002	PM Goh replied to PM Mahathir. He noted that since Malaysia wanted to discontinue the package approach, Singapore would have to deal with water and the other issues on their stand-alone merits and no longer as a package.
16–17 October 2002	Senior officials from both sides met in Johor Bahru to discuss the water issue. But Malaysia wanted only one aspect of water discussed – the price of current raw water. Singapore reiterated that Malaysia had lost its right to the review, but it would agree if Malaysia agreed also to discuss the supply of future water. Singapore also asked Malaysia to explain how it had arrived at the price of RM 6.25 it was asking for. It said that going by the terms of the Water Agreements, any review would result in a price of not more than 12 sen in 2002. Malaysia could not provide a satisfactory explanation.

Source: MICA (2003, Annex C, pp. 82–83).
Note: The version available at http://app.mfa.gov.sg/data/2006/press/water/event.htm is somewhat different to the printed version in the wording of the events and also in the detail in which it describes the several events. The version available on-line also includes comments on two meetings on 14 and 25 March 2002, which are not included in the printed version and therefore are not included in this table.

China's Legal System for Water Management: Basic Challenges and Policy Recommendations

PENG SHUGANG

Complaint Reception Office of Shanghai Municipal Government, Shanghai, PR China

ABSTRACT *During the period 1978–2008, China's legal system for water management experienced a positive evolution process, but all the management measures were heavily reliant on administrative regulation with limited application of a market mechanism, which was exacerbated by limited public participation and offset by distorted incentives on the part of the government officials in law enforcement. Empirical study reveals that the current legal reform is inadequate to redress the challenges in water management. An integrated water management system based on a market mechanism, public participation, and a sensible incentive for government officials is advisable to resolve the looming water crisis in China.*

Water Policy in China: An Overview

Brief Introduction to China's Rivers and Lakes

Ever since the opening and reform policy was adopted by the ruling party in 1978, China has achieved an average 10% annual gross domestic product growth for three decades, against the backdrop of decreasing per capita water quantity and deteriorating water quality. Water, as one of the most essential natural resources for human survival and development, is a case in point on denoting how natural resources have been over-exploited in an unsustainable way in China. This paper reviews the whole evolution process of China's water policy formulation during the period 1978–2008 from three perspectives: the water management system, water quantity management, and water quality management. It starts with a description of the status quo and policy response from the legislature and the government, followed by policy recommendations based on diagnostic analysis of the loopholes in the water management legal system.

There are seven major river basins in China: Yangtze, Yellow, Pearl, Heilong, Liao, Hai-luan, and Huai Rivers. Precipitation across China is unevenly distributed. Rainfall in the northern part of the country is inadequate in most years, whereas the southern part of China is endowed with adequate precipitation (Table 1). The uneven distribution of water resources, coupled with a huge demand for water because of the large

Table 1. Water resource distribution in North and South China

National values	Northern five major watersheds	Southern four major watersheds
Water resources (%)	19	81
Population (%)	46.5	53.5
Per capita water resources (m^3)	1,127	3,381
Gross domestic product (%)	45.2	54.8
Cultivated land (%)	65.3	34.7

Sources: Ministry of Commerce (2002), quoted in US Department of Commerce, International Trade Administration (2005).

population size, puts the northern part of China in danger of running into a severe water shortage crisis.

Policy Response, 1978–2008

Based on the government's perception of water as a natural resource, the development of China's water management can be subdivided into three stages:

- *1978–1987: The application of conventional wisdom with limited involvement of the rule of law.* During this period, the government adopted an approach of 'tapping the stones when crossing the river'. There was no law in place to govern water quantity management, many government authorities could interfere in water management and no single government authority was directly responsible for water affairs. As regards water quality management, the Prevention and Control of Water Pollution Act was promulgated in this period (Standing Committee of National People's Congress (SCNPC), 1984; hereafter the Water Pollution Act 1984).
- *1988–2001: Water as an instrument to maximize economic benefits.* The framework of water policy began to take shape during this period. In 1988, the state legislature promulgated the Water Act (SCNPC, 1988), which stipulates that water resource authorities at various levels of governments are responsible for water management. In order to stop water quality from deteriorating even further, the State legislature amended the Water Pollution Act 1984 in 1996 (SCNPC, 1996; hereafter the Water Pollution Amendment Act 1996). During this period, more attention was paid to the efficient use of water to facilitate economic growth.
- *2002 to date: Water as an integral part of the natural resources for sustainable development.* Overemphasis by the government on the efficient utilization of water resources has been brought about by rapid deterioration of water quality and fast depletion of this natural resource during more than a decade. Thus, the government has had to take a serious attitude toward sustainable preservation and the utilization of water resources. In 2002, a new Water Act (SCNPC, 2002) was promulgated to replace the Water Act 1988. The government began to take a more holistic attitude toward water management by trying to achieve a balance between economic growth and preservation of the environment. In 2008, the State legislature amended the Water Pollution Act 1984 for the second time (SCNPC, 2008; hereafter the Water Pollution Amendment Act 2008).

Table 2. Selected laws and regulations in China's water policy

Category	Title	Rule-making Authority (Authorities)	Year of promulgation	Year(s) of revision
National-level laws and regulations	Water Law	State Legislature	1988	2002
	Environmental Protection Act		1999	
	Water Pollution Prevention Act		1984	1996, 2008
	Environmental Impact Assessment Act		2002	
	Urban and Rural Area Planning Act		2008	
	Clean Production Promotion Act		2002	
	Water and Soil Conservation Act		1991	
	Circular Economy Promotion Act		2008	
	Regulations on Water-Withdrawal Permit and Collection of Water Resource Fee	State Council	2006	
	Regulations on Municipal Water Supply		1994	
Ministry-level rules	Rules on Hygiene Supervision of Domestic Potable Water	Ministry of Construction, Ministry of Health	1996	
	Provisional Regulations on Water Quantity Allocation	Ministry of Water Resource	2008	
	Rules on Quality Management of Municipal Water Supply	Ministry of Construction	1999	2004, 2007
	Rules on Pricing Management of Urban Water Supply	National Development and Reform Commission, Ministry of Construction	1998	
	Rules on Qualification Management of Municipal Water Supply Enterprises	Ministry of Construction	1993	
	Rules on Water Withdrawal Permission	Ministry of Water Resource	1994	2008

During the past three decades, the State legislature and government have promulgated a series of laws and regulations to regulate water management (Table 2). They cover almost every aspect of water management from strategic planning and water quantity management to water quality control. Thus, a legal system for water management has come into shape within three decades.

Institutional Framework for Water Management

The Ministry of Water Resources has been playing an increasingly important role in water management. According to the Water Act 2002, the power of water management in China is shared by the Ministry of Water Resources and local governments. The Ministry is responsible for overall water management across the country; seven River Basin Commissions are responsible for the daily administration of water management within their scope of power delegated by the Ministry of Water Resources. One of the most important revisions in the Water Act 2002 is that the new law attaches more importance to strategic planning, rather than passively reacting to the changing situation. As a result of legal reform, the power of water management has been increasing centralized in the hands of the Ministry of Water Resources.

To the author's great disappointment, the model of 'nine dragons administering water' still remains intact even up to now. There are eight other authorities under the State Council involved in water policy (Tables 2 and 3), and this has resulted in a fragmented water management system permeated with an intense power struggle among the authorities. For example, according to the Regulations on Municipal Water Supply (State Council, 1994), municipal construction authorities at various levels of government are responsible for supervising the urban water supply. Nevertheless, the health authorities are empowered to regulate the quality of drinking water according to the Communicable Disease Prevention Act (SCNPC, 1989a). In order to settle the power dispute on the supervision of water quality, the Ministry of Construction and the Ministry of Health achieved a compromise by stipulating that the Ministry of Health is responsible for the nationwide quality supervision of drinking water, while the Ministry of Construction is responsible for the hygiene management of urban drinking water (Ministry of Construction and Ministry of Health, 1996). Due to a lack of coordination, the legislature and the government have failed to distribute powers among different authorities according to real needs. As a result, the Ministry of Water Resources has been marginalized significantly and inefficiency in water management becomes inevitable.

Apart from the trouble in the horizontal distribution of power in water management, China is also plagued by problems in its vertical distribution of power among various levels of government. As indicated in Figure 1, the local water resources authorities and environmental protection agencies only report to their corresponding levels of government, and the Ministry of Water Resources and the Ministry of Environmental Protection are only entitled to 'guide' them on certain issues. If the local authorities violate the 'guidance' from the Ministry of Water Resources or the Ministry of Environmental Protection, the two ministries have no coercive power to force them to abide by the instructions.

On the other hand, the power of water management has been highly concentrated at various levels of government rather than being delegated to the competent authorities (Figure 1). According to Articles 29 and 39 of the Environmental Protection Act (SCNPC, 1989b), only above-county-level governments are empowered to order the

Table 3. Water management – government agencies in China

Department	Scope of water administration responsibilities	Major functions
Ministry of Water Resources	Surface and groundwater management, river basin management, flood control, water and soil conservation	Planning of water development and conservation, flood control, water and soil conservation, water function zoning, medium- and long-term master plan for water demand and supply, strategic plan for water quantity allocation
Ministry of Environmental Protection	Prevention and treatment of water pollution	Water environmental protection and water environmental function regionalization/zoning to establish national water environmental quality standards and national pollutant discharge standards
Ministry of Construction	Urban and industrial use, urban water supply and drainage	Planning, construction and management of water supply projects and drainage and sewage disposal projects
Ministry of Agriculture	Water use for irrigation and fishery industry, protection of aquatic environment	Non-point source pollution control, protection of fishery water environment and aquatic environment conservation
State Forestry Bureau	Water resources conservation	Forest protection and management for protecting watershed ecology and water resources
State Electricity Regulation Commission	Hydro-power development	Construction and management of large- and medium-scale hydro-power projects
State Development and Reform Commission	Participation in the planning of water resource development and conservancy of ecosystem	Planning of water resource development, allocation of factors of production and conservancy of ecosystem, coordination and planning of agricultural, forestry and water resources management
Ministry of Communication	Water pollution control on shipping in inland rivers and lakes	Pollution control and management of inland navigation
Ministry of Health	Supervision and management of environmental hygiene to protect human health	Rule-making and supervision to ensure the quality of potable water

Source: Feng *et al.* (2006), with modifications made by the author according to latest developments.

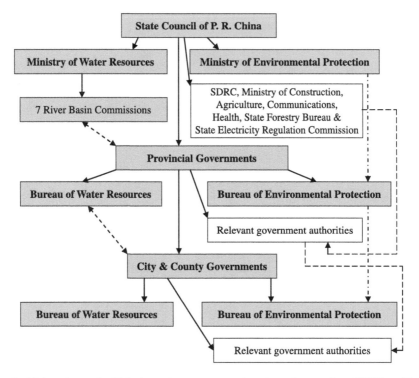

Figure 1. Major players in China's water management system. *Source*: Lee (2006), with some modifications made by the author.

forced closure of polluting enterprises in cases of serious pollution, yet environmental protection agencies are left with no other options but to report to the same level of government for a final decision. On the other hand, under the current taxation system, the state-owned enterprises are obliged to hand in their corporate income tax to the respective levels of governments which enjoy the ownership of state-owned enterprises. Under the current corporate income tax arrangement, local governments have a strong motivation to protect locally owned state-owned enterprises, while enterprises owned by the central government are always exempt from local law enforcement because local governments are not empowered to impose punitive measures against their illegal acts. To conclude, only certain levels of government can have the final say in law enforcement on water pollution issues.

The negative impacts of the flaws within the current legal system have been amplified by lax administrative law enforcement. Under the current personnel system, the ability to achieve economic growth has been the major, and on many occasions almost the only, criterion for the promotion of government officials; cultural or social achievements were largely ignored in the appraisal of the merits of government officials. The reward system induces officials to endeavour to achieve fast gross domestic product growth for the sake of their career prospects during their tenures, even at the huge cost of serious water pollution.

The fragmented water management structure has made the cost of coordination among different authorities extremely taxing and expensive and highly inefficient. Fragmentation of the water management system makes coherent policy formulation and implementation more difficult, thus the whole country is subject to unsustainable water use and worsening water pollution. In fact, the Asian Development Bank (2005) has identified the lack of a neutral water sector apex body as a key problem for China's future water sector reform.

Water Quantity Management

Current Situation of Water Quantity

Though the population growth rate of China decreased significantly from 1.2% in 1978 to 0.52% in 2007, with an annual average natural growth rate of only 1.06%, a net increase of 358.7 million people was witnessed between 1978 and 2007. What is worse, China's population is projected to reach 1.6 billion by 2050. On the other hand, available water resources keep a downward trend measured either by the total amount or on a per capita basis (Table 4). It should be noted that both surface water and groundwater decreased rapidly during the past eight years. This indicates that people in arid northern China may face an even more pessimistic future.

In sharp contrast, water use increased quickly during the same period (Table 5). The amount of water used in the industrial sector and daily consumption increased steadily, while water use in the agricultural sector decreased only slightly, and per capita water consumption kept an upward trend in the meantime. Increasing water consumption and population size, coupled with decreasing water resources available, poses a great threat to China's water security.

Basic Approaches in Water Quantity Management

In order to handle effectively the challenges in water quantity management, China has put forward the following measures.

Quantity control and quota management. According to Article 47 of the Water Act 2002, provincial-level regulating authorities are obliged to set water consumption quotas for the relevant industrial sectors. The quota should be made public by the provincial governments and submitted to the Ministry of Water Resources for record. The operation of the water quantity-control mechanism is based on the quota system. On 1 February 2008, Provisional Rules on Water Quantity Allocation was promulgated (Ministry of

Table 4. China's water resources, 2000–2007

Year	Total amount of water resources (100 million m^3)	Surface water	Groundwater resources	Duplicated measurement between surface water and groundwater	Per capita water resources (m^3/person)
2000	27,700.8	26,561.9	8,501.9	7,363.0	2,193.9
2001	26,867.8	25,933.4	8,390.1	7,455.7	2,112.5
2002	28,261.3	27,243.3	8,697.2	7,679.2	2,207.2
2003	27,460.2	26,250.7	8,299.3	7,089.9	2,131.3
2004	24,129.6	23,126.4	7,436.3	6,433.1	1,856.3
2005	28,053.1	26,982.4	8,091.1	7,020.4	2,151.8
2006	25,330.1	24,358.1	7,642.9	6,670.8	1,932.1
2007	25,255.0	24,242.0	7,617.0	6,604.0	1,921.3

Sources: National Statistics Bureau (2008); Ministry of Water Resources (2008d).

Table 5. Water consumption in China, 2000–2007.

Year	Water use (100 million m^3)	Agriculture	Industry	Consumption	Ecological protection	Per capita water use (m^3/person)
2000	5,497.6	3,783.5	1,139.1	574.9	n.a.	435.4
2001	5,567.4	3,825.7	1,141.8	599.9	n.a.	437.7
2002	5,497.3	3,736.2	1,142.4	618.7	n.a.	429.3
2003	5,320.4	3,432.8	1,177.2	630.9	79.5	412.9
2004	5,547.8	3,585.7	1,228.9	651.2	82.0	428.0
2005	5,633.0	3,580.0	1,285.2	675.1	92.7	432.1
2006	5,795.0	3,664.4	1,343.8	693.8	93.0	442.0
2007	5,819.0	3,602.0	1,402.4	709.9	104.7	440.4

Note: n.a., not available. *Sources*: National Statistics Bureau (2008); Ministry of Water Resources (2008d).

Water Resources, 2008a). According to this regulation, the quantity of water supply within a given jurisdictional area is to be fixed by quota in the years to come and monitored by the Ministry of Water Resources.

Taking the approach adopted in Guangdong Province as an example (Bureau of Water Resources of Guangdong Province, 2007), the calculation of the quota is based on the following formulas:

$$\text{Quota for industrial water use} = \text{total amount of output} \times \text{amount of water withdrawal for the specific product per unit} \times (1 + \text{ratio of water reuse}) \quad (1)$$

$$\text{Quota for agricultural water} = \text{total amount of acreage} \times \text{amount of water withdrawal/year/acreage} \quad (2)$$

Note that during recent years, the implementation of quota restrictions has been increasingly stringent, and the amount of water use exceeding the prescribed quota is subject to price increase on a progressive basis according to the Water Act 2002.

According to the Regulations on Water-withdrawal Permit (Ministry of Water Resources, 2008c), any organization or person who intends to access water should obtain prior permission from the water resource authorities, whereby the total amount of water subject to the permit should not exceed the water resources available in the given jurisdiction area. Thus, the water quota system and water-withdrawal permit system complement each other, and they have become instrumental for water management authorities to achieve effective water quantity control.

Water pricing mechanism. Any organization or person who draws water from rivers, lakes or underground should make a payment to the water resource authorities according to price standards prescribed by provincial governments. The water revenue should be shared by the central and respective local governments. In addition, a progressive water resource pricing system is adopted to counterattack water-withdrawal which goes beyond quota restrictions.

The heavily subsidized urban water supply has been a major reason for discouraging people to conserve water. According to Article 26 of the Regulations on Municipal Water Supply (State Council, 1994), the price mechanism of municipal water supply should ensure water supply companies cover their costs and achieve a reasonable profit, and the price level of the water supply is determined by the provincial governments. The Rules on Pricing Management of Urban Water Supply, promulgated by the National Development and Planning Commission and Ministry of Construction (1998) (see Table 2) stipulate that water pricing should abide by the following principles: (1) it covers operation and maintenance, depreciation, and interest costs; (2) it allows an 8–10% return on the net value of fixed assets; and (3) it is appropriate to local characteristics and social affordability. Due to various considerations, such as social affordability and the need to push down the cost of water use in production for local producers, local governments are reluctant to adopt a water pricing system that abides by the cost-recovery principle which is reflected in the above-mentioned rule for water pricing.

The available facts have proved that though water tariffs in China have been increasing constantly since 1978, it still lags behind the actual cost. For example, in Beijing, the capital city of China with severe water scarcity, the price for water supply increased from 0.80 RMB in 1997 to 3.70 RMB per cubic metre in 2004. Although the price level of water in Beijing is currently the highest among all cities in China, water and sewerage in the city still remain subsidized, even in strictly financial terms (The World Bank, 2007).

Preferential treatment for the pricing of agricultural water has been another hurdle to effective water conservation in China. Due to a consideration of water affordability for rural residents, water price for the agricultural sector has been set at an artificially low level. A government regulation, the Administrative Instructions on Water Pricing Reform in the Agricultural Sector (State Planning Commission, 2001), reveals that since the pricing of agricultural water is based on the acreage of cultivated farmland instead of the actual amount of water used, there is no incentive for farmers to save water. According to the Rules on Pricing of Water Supply from Water Conservation Projects (State Development and Reform Commission and Ministry of Water Resources, 2003), the agricultural sector enjoys preferential treatment in water use in terms of pricing. The pricing of water for agricultural use is determined by the cost of the water supply, while the pricing of water for use in the non-agricultural sector is based on the cost of water supply plus tax plus profit. The tax and profit rate equals more than 2–3% of the prevalent long-term interest rate prescribed by commercial banks.

China's preferential treatment for agricultural water has proved to be a disincentive for investment in agricultural infrastructure, which in turn results in a waste of water in agricultural production. China's stagnant growth in effective irrigation area, namely, the cultivated land equipped with irrigation facilities, has contributed to a waste of water in the agricultural sector. As Figure 2 indicates, only around 50% of all the cultivated land was equipped with an irrigation system between 1978 and 2000; when the total area of cultivated farmland increased dramatically in 1999 and 2000, no progress was realized in upgrading the irrigation system, so the increased area of farmland without an irrigation system calls for more water to be allocated to agricultural production. This conclusion coincides with the data shown in Table 5, which indicate that water use in the agricultural sector accounted for a lion's share of total water use during 2000–2007, and the reduction achieved in agricultural water use was unimpressive during the same period.

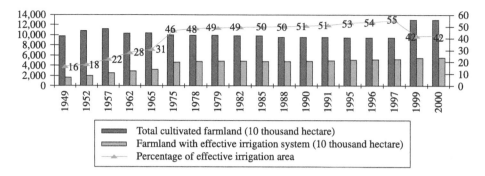

Figure 2. China's effective irrigation area during 1949–2000. *Source*: Han & Zhao (2004).

Water right transaction mechanism. One of the basic flaws in the Water Act 1988 has been that the government is heavily dependent on administrative measures to regulate water affairs, but simply relying on government intervention may lead to government failure. A regulation promulgated by the central government has paved the way to resolve this problem. According to Article 27 of the Regulations on Water-withdrawal Permit & Collection of Water Resource Fee (State Council, 2006), every entity or individual has the right to transfer a water-withdrawal right within the water-withdrawal quota to other entities or individuals within the effective period acknowledged by the permit, provided that the above-mentioned water quota has been conserved as a result of technological improvement or due to adjustment of product or industrial structure.

The transfer of a water withdrawal right will undoubtedly facilitate efficient water use, which will exert a long-term positive impact on China's water management. On 24 November 2000, Dongyang and Yiwu cities in Zhejiang Province reached an agreement on the transfer of the water right. According to the agreement, Yiwu City ensured its permanent right to claim 500 million m³ of water annually from Hengjin Reservoir in Dongyang City for a one-time payment of 200 million RMB (equivalent to US$29.3 million), subject to another annual payment for the water resource fee according to the prevalent price level set by the provincial government. This has been the first case of a transfer of a water right in China (Hu & Wang, 2001).

The Dongyang-Yiwu water right transaction indicates that China has succeeded in allocating water resources by means of market mechanisms instead of by direct administrative intervention. Given the fact that the nationwide water allocation system came into operation in 2008, and the water-quota system will be applied indiscriminately to all entities and citizens across China, more transactions of water rights are likely to be observed in the years to come. It is reported that the South–North Water Transfer Project (midline part) is scheduled to be completed in 2014. This project, which runs across many provinces and cities from central China to Beijing, will undoubtedly have a great impact on water quantity allocation in many provinces and municipalities. It is urgent for the government to design an effective water right transaction mechanism to rationalize the water supply, otherwise it is highly probably that the current predatory water consumption model will be translated into another water crisis in Southern China (Ministry of Water Resources, 2008b).

In summary, the government has been relying heavily on administrative intervention in water quantity management; since the pricing system still has a long way to go before it conforms to the 'cost-recovery' principle, and the water rights transaction mechanism is still in its infancy, more efforts are needed to introduce market mechanisms into water quantity management.

Water Quality Management

Status Quo of Water Quality Management

China's rapid industrialization has been accompanied by increasing wastewater discharge and serious water pollution. As Table 6 indicates, the total amount of discharged wastewater is increasing dramatically with the growth of China's economy. Though the pollution, as indicated by the total amount of chemical oxygen demand, decreased slightly due to improvements in the treatment of industrial pollution, organic pollution from non-industrial sectors as indicated by chemical oxygen demand has increased very quickly. In comparison, pollution from ammonia nitrogen also increased in the same period due to significant emissions from non-industrial sectors. Since a large share of ammonia nitrogen from non-industrial sectors comes from untreated human and animal excrement and the increasing use of fertilizer in agriculture (Figure 3), the removal of ammonia nitrogen from water can be even more challenging due to the fact that fertilizer consumption in agriculture is non-point pollution by nature whereas control and treatment of non-point pollution appears to be more inefficient and difficult in comparison with abatement of point pollution from a technical point of view.

No obvious improvement is witnessed between 1980 and 2001 in terms of water pollution treatment. As indicated in Figure 4, organic water pollutant emissions measured by biochemical oxygen demand per day increased steadily during 1980–1990, followed with even more significant growth during 1991–2001. On the other hand, organic water pollution emission fluctuated within the range 0.13–0.15 kg/day per worker, which means there was no significant improvement in terms of pollution treatment during the whole period.

Throughout 2000–2007, the quality of river water kept a deteriorating trend. As indicated in Figure 5, around 40% of the total river length which is subject to appraisal by the Ministry of Water Resources, has been seriously polluted (with water quality in and above Grade IV). According to official reports from the Ministry of Environmental Protection (Table 7), among the 28 major lakes (reservoirs) in China monitored by environmental protection agencies in 2007, only two lakes (reservoirs) met a Grade II water quality standard; medium-level pollution was witnessed in six lakes (reservoirs); the other 20 lakes (reservoirs) were seriously polluted. During 2003–2007, water pollution increased steadily.

To conclude, China has been plagued by both a high level of water pollutant emissions and poor improvement in water pollution treatment. The country's inadequate supply of water resources has been exacerbated by deteriorating water quality nationwide.

Policy Response During 1984–2008

The National People's Congress adopted the Prevention and Control of Water Pollution Act in 1984 (Act 1984). This law was later amended twice in 1996 and 2008. According to the amendments, the following approaches are adopted to counteract water pollution.

Imposition of a quota on pollutant discharge. It should be noted that the Chinese government only began to take this issue seriously very recently.

Table 6. Discharge of wastewater and major water pollutants in China

Year	Wastewater discharge (100s million tons)			Chemical oxygen demand (10,000s tons)			Ammonia nitrogen (10,000s tons)		
	Total	Industrial	Non-industrial	Total	Industrial	Non-industrial	Total	Industrial	Non-industrial
1998	395.3	200.5	194.8	1,495.6	880.6	695			
1999	401.1	197.3	203.8	1,388.9	691.7	697.2			
2000	415.2	194.2	220.9	1,445	704.5	740.5			
2001	432.9	202.6	230.3	1,404.8	607.5	797.3	125.2	41.3	83.9
2002	439.5	207.2	232.3	1,366.9	584	782.9	128.8	42.1	86.7
2003	460	212.4	247.6	1,333.6	511.9	821.7	129.7	40.4	89.3
2004	482.4	221.1	261.3	1,339.2	509.7	829.5	133	42.2	90.8
2005	524.5	243.1	281.4	1,414.2	554.8	859.4	149.8	52.5	97.3
2006	536.8	240.2	296.6	1,428.2	542.3	885.9	141.3	42.5	99.8
2007	556.7	246.5	310.2	1,381.8			132.3	34	98.3

Sources: Ministry of Environmental Protection (2002–2008).

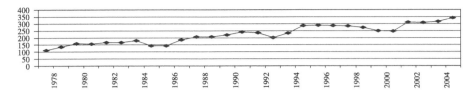

Figure 3. Fertilizer consumption, 1978–2005 (kg per hectare of arable land). *Source*: The World Bank (2008).

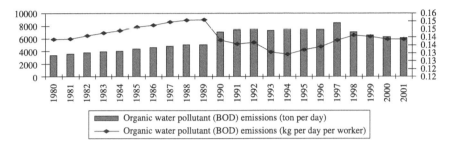

Figure 4. Organic water pollutant (biochemical oxygen demand—BOD) emissions, 1980–2001. *Source*: The World Bank (2008).

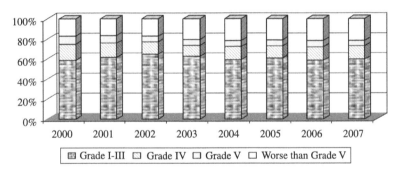

Figure 5. Water quality of rivers measured by the percentage of total river length subject to appraisal by the Ministry of Water Resources, 2000–2007. *Sources*: Ministry of Water Resources (2001–2008).

Article 16 of the Amendment Act 1996 stipulates that provincial and central governments *may* adopt control measures on pollutant discharge in certain highly polluted reaches and lake areas. Since the central and local governments were given discretionary power in this respect, the various governments made no essential effort to reduce the discharge of pollutants into water bodies. It was only until 2007 that central government put forward its implementation plan on a reduction of pollutant emissions according to the requirement of the 11th Five-Year Master Plan (2006–2010). According to this implementation plan (State Council, 2009), during the 11th Five-Year Master Plan period, the major pollutants are projected to reduce by 10%, chemical oxygen demand be reduced

Table 7 Water quality of major lakes and reservoirs, 2007

| Water body | Number | Water quality grade | | | | | |
		I	II	III	IV	V	>V
Three Lakes	3	0	0	0	0	1	2
Big freshwater lake	10	0	0	2	4	1	3
Urban lake	5	0	0	1	0	0	4
Big reservoir	10	0	2	3	0	3	2
Total	28	0	2	6	4	5	11
Percentage in 2007		0	7.1	21.4	14.3	17.9	39.3
Percentage in 2006		0	7.0	22.0	4.0	19.0	48.0
Percentage in 2005		0	7.0	21.0	11.0	18.0	43.0
Percentage in 2004		0	7.5	18.5	14.8	22.2	37.0
Percentage in 2003		0	3.6	21.4	25.0	14.3	35.7

Note: In 2004 and 2006, only nine big freshwater lakes were included in the reports. *Source*: Ministry of Environmental Protection (2004–2008).

from 14.14 million tons in 2005 to 12.73 million tons in 2010, and the percentage of treated wastewater should be no less than 70% before final discharge into water bodies.

In the Amendment Act 2008, provincial-level governments are required to reduce and control the total amount of water pollutants according to the prescriptions of the State Council, the authorized discharge quota of water pollutants is to be distributed to county-level government, followed by another round of further distribution to the pollutant-discharging entities. If in any given jurisdiction area the discharged pollutants exceed the stipulated quota, the relevant government is obliged to suspend issuing permits to new construction projects. Furthermore, any county or enterprise that has failed to abide by the quota should be made public.

Pollutant discharge permit. According to the Amendment Act 2008, any entity that directly or indirectly discharges industrial or medical sewage should get prior discharge approval from the government. This is the first time the law stipulates that permission from the government is a precondition for wastewater discharge.

Loopholes in the old law and lack of regulation led to pervasive water pollution across the country before 2008. According to Article 14 of the Water Pollution Prevention Amendment Act 1996, entities that directly or indirectly discharge pollutants into water bodies should report their pollutant discharge facilities, as well as the type, amount and density of the discharged pollutants, to the pertinent environmental protection agencies, and provide basic technical materials on water pollution treatment. Pollutant discharge fees are levied according to the report from the polluters. Since the registration process is a confirmation of pollution discharge, polluters are induced to understate or even avoid reporting to the environmental agencies the details of their pollution (Luo, 2008). Given the inadequate human resources available for law enforcement in water pollution control, it is easy for polluters to take advantage of the loopholes in their reporting to maximize their benefits. The 'report-after-pollute' management model has been an important contributor to the deteriorating water pollution in China during the past few years.

The lack of pollutant discharge permit legislation will jeopardize the government's new effort to achieve quota control in pollutant emission. According to Article 20 of the

Amendment Act 2008, the implementation rules for pollutant discharge approval should be prescribed by the State Council in this respect, but currently this regulation is still under consideration by the State Council, despite calls from the public to accelerate the legislation process. If the pollution discharge permit system is not put into place, the imposition of quota control on pollution discharge will be impracticable, since a government within a given jurisdiction area bears no legal liability for surpassing the quota according to the Amendment Act 2008. Based on past experience, it could be argued that the effectiveness of the new water pollution prevention act will be offset significantly if the regulation on pollutant discharge permit management cannot be promulgated on a timely basis.

Merit-appraisal mechanism of chief government officials in water pollution control. In the Amendment Act 2008, the law stipulates that a review system should be adopted when appraising the performance of local chief officials during their tenure, which means local politicians will be placed in a disadvantaged position if they fail to pass the appraisal on their merits of water preservation and pollution control, and this will cast a cloud on their career prospects.

It should be noted that currently there are no rules available to accommodate the merits of government officials in environmental protection into the government personnel system, so whether this article can be effectively implemented depends largely on whether there will be adjunct rules available to put this policy objective into effect.

Imposition of a sewage discharge fee. According to Act 1984 and the Amendment Act 1996, an enterprise may continue to operate even though it discharges water pollutants well above the permitted discharge standard, as long as a discharge fee is paid to the government within a given period. Since a discharge fee for excessive pollution only consists of a small share of the total cost needed for pollution treatment, the cost of water pollution is thus externalized, and polluters have benefited markedly from excessive pollution. The new law has significantly increased the cost of excessive pollution discharge to deter polluters. According to the Amendment Act 2008, any enterprise that fails to abide by the pollution discharge standard will be fined two to five times the discharge fee, and runs the risk of forced closure by the government if it fails to reach the discharge standard within one year. As pointed out above, environmental protection authorities have not been empowered to command forced closure of polluting entities in cases of serious pollution up to the present, so how this article will be implemented effectively remains to be seen.

To sum up, the Chinese government only showed its will to tackle the problem of water pollution very recently, and the stipulations for water pollution control have been more rigorous in recent years. But based on a deliberation of the whole structure of laws and regulations, it is evident that more implementation rules are needed to put the articles into effect. Apart from this challenge, the current laws attach too much importance to government intervention to tackle water pollution, while how to make the best use of market mechanisms in water pollution control has been systematically ignored in the legislation process.

As regards utilization of market mechanisms in pollution control, one of the significant flaws of the Water Pollution Prevention Act 2008 is that it fails to make use of tradable pollution rights. If enterprises can trade their pollution rights with each other, then those enterprises that bear a lower cost by a reduction in pollution can sell their pollution quota to those that have to bear a higher cost if they are supposed to abide by the pollution control rules. Thus, both sellers and buyers in the market can benefit from the transaction of pollution rights. Since there is a potential to get economic benefit from pollution reduction, a market for the exchange of pollution rights will arise. Though the transaction of pollution rights may not necessarily reduce the total amount of pollution within a given jurisdiction/area, the government can still reduce overall pollution by lowering the pollution quota for a given area (Beder, 2001). Fortunately, despite there being no law in place to govern the pollution rights transaction mechanism, pilot programmes have been implemented throughout Zhejiang, Jiangsu provinces and Tianjin city up to now (Deng, 2009).

Conclusions: Policy Recommendations

Currently, the per capita water resources available for Chinese citizens are less than one-quarter the average level around the world; 400 out of more than 600 cities in China are encountering the challenge of water shortages, and the daily gap between water supply and demand reaches 180 million tons for the 400 cities. This is the sober reality faced by the Chinese government and the public, thus it is advisable for the government to take more effective policy measures to resolve the current water crisis.

Establish an Integrated Water Management System

The facts cited in this article prove that China's water policy during 1980–2007 was not successful, and one of the root causes of ineffective water policy lies in the disintegration of the water management system. In view of the current problems, China needs to establish a water management system by consolidating power within the Ministry of Water Resources and the Ministry of Environmental Protection, and the following measures are necessary to achieve this objective.

Affairs pertinent to water quantity management. Responsibility for affairs pertinent to water quantity management should be conferred entirely on the Ministry of Water Resources. Responsibility for the following affairs conferred on other authorities should be transferred to the Ministry of Water Resources as soon as possible: irrigation water management (from the Ministry of Agriculture); planning, construction and the management of water supply projects, as well as drainage and sewage disposal projects (from the Ministry of Construction); and supervision and management of drinking water standard (from the Ministry of Construction and the Ministry of Health).

Environmental protection authorities. Environmental protection authorities should be empowered adequately to meet the demand for more stringent and responsive law enforcement. Currently, China needs to accelerate its legislation progress in drafting the Regulations on Environmental Impact Assessment for Urban & Rural Planning, thus mitigating the negative impact of government zoning on water pollution. On the other hand, the Environmental Protection Act 1994 needs to be amended soon to make the power conferred upon the environmental protection agencies match with their legal responsibility. Powers such as the decision-making power of forced closure of seriously polluting enterprises will be critical instruments for effective law enforcement, otherwise the authorities will be 'tigers without teeth'.

Apply More Market Mechanisms to Provide Disincentives for Over-Consumption of Water and Increasing Water Pollution

Up to now, the current legal system has attached much importance to the role of administrative intervention in water management. The allocation of water resources and treatment of polluted water are heavily dependent on government monopoly and administrative supervision. The current legal system is heavily dependent on traditional approaches of government intervention, such as quotas for water allocation in different jurisdiction areas as well as quotas for the amount of water withdrawal by water users, to tackle the challenges in water management. Since it is technically difficult to stipulate the liability of over-withdrawal of water resources, and there are no liability clauses in the current laws and regulations to regulate the above-mentioned behaviour, the constraints faced by administrative intervention are obvious and may not necessarily be effective in tackling the problems.

The cost-recovery principle should be reflected in the pricing mechanism. One of the challenges staring at the government in promoting water conservation comes from the long-standing negative effect of an artificially low price level. If the government ignores the urgent need to discourage over-consumption of water by catering to the needs of vested interests (a lower cost for production) and social equity (care for the poor), the water crisis is likely to be unmanageable in a short time. Though the author does not deny the fact that affordability is an important issue for water pricing, especially for the poor and underprivileged rural residents, the government can provide financial assistance to them on a strict means tested basis instead of artificially lowering the price below the actual cost of water supply. A gradual approach to pricing reform is needed to reflect the real cost of water supply and to deter the over-consumption of water.

As regards water quantity management, it is essential to clarify water property rights for different stakeholders within the current legal system. Once the amount of allocated water has been designated for a given jurisdictional area, if there is no market available to satisfy the need for the transaction of water rights, water users who have adequate resources will be prone to over-consumption once government supervision is not in place, while water users who reside in areas without adequate rainfall will suffer from chronic water shortages (Li & Luo, 2006).

With regard to water quality management, rules for the transaction of water pollution rights are indispensable for pollution control. It is necessary for the legislature and government to design rules on tradable pollution rights, otherwise it is impracticable for the government to ensure effective pollution control of the total amount of pollutants discharged to water bodies.

Facilitate Public Participation to Enhance Water Governance

Without public involvement in water management, it is difficult to hold the government accountable to the public interest given that government officials may run for office with one eye on their own interests such as higher tax revenues and more chances of career promotion at the cost of public welfare. In China's context, the following measures are needed in this respect.

Encourage public participation in water management. According to the Regulations on Public Access to Government Information (State Council, 2007), the public are entitled to know the details about law enforcement in environmental protection as well as other water management issues which are pertinent to their own interest. Support from the public will be instrumental for the government's effort to control water pollution and improve water management. Apart from that, the public are entitled to voice their concerns according to the Provisional Rules on Public Participation in Environmental Impact Assessment (State Environmental Protection Administration, 2006), whereas legal support is not available for public involvement in other aspects of water management. In view of the current situation, it is advisable for the state legislature to promulgate the Administrative Procedure Act. Thus, government behaviour will be put under close scrutiny, whereby public participation may exert great influence in aspects of public hearing, pollution control, and law enforcement as well as decision-making in water management. It should be noted that there is still room for local authorities to enhance public participation. For example, the Rules on Administrative Procedure of Hunan Province (Hunan Provincial Government, 2008) were adopted by the Hunan provincial government last year. This is the first time that a provincial-level government has endeavoured to regulate government behaviour and enhance good governance. Other provincial legislatures and governments may follow suit to encourage public participation in water management.

Empower the public to make public interest litigations against water pollution. This is one of the most practical approaches to ensure the participation rights of the public. To the author's disappointment, even the latest amendment to the Civil Procedure Act 2007 (SCNPC, 2007) has not granted the public, non-government organizations (NGOs), or the procuratorial authorities the right to lodge public interest litigations against water polluters, and public interest litigations initiated by individuals or NGOs are usually rejected by the Chinese court system.

Reform the Current Promotion System for Government Officials to Boost Sustainable Water Management

China's promotion system for government officials, in which the career prospects of local officials is largely dependent on their capacity to achieve economic growth within their tenure of office, has pushed government officials to prioritize economic growth to the extreme, even at the cost of negative social and cultural consequences. In order to inspire the officials to strive for the long-term benefits of society based on a comprehensive cost–benefit analysis of policy impacts, it is high time that the ruling party modified its current promotion system for government officials, otherwise the effort of achieving efficient water management might be ruined by the distorted motivations of officials at various levels of government.

References

Asian Development Bank (2005) *Charting Change: The Impact of ADB's Water For All Policy on Investment, Project Design and Sector Reform* (Manila: Asian Development Bank).

Beder, S. (2001) Trading the earth: the politics behind tradeable pollution rights, *Environmental Liability*, 9, pp. 152–160.

Bureau of Water Resources of Guangdong Province (2007) *Quota for Water Consumption in Guangdong Province (Provisional Scheme)*. Available at http://www.gdwater gov.cn/export/download/gonggongxiazaiwenjianku/1170836370750.doc (accessed 11 April 2009) [in Chinese].

Deng, L. (2009) Pilot scheme for pollution rights transaction in Zhejiang get approval, *21st Century Economic Report (Guangzhou)*, 14 April [in Chinese].

Feng, Y., Daming, H. & Kinne, B. (2006) Water Resources Administration Institution in China, *Water Policy*, 8, pp. 291–301.

Han, H. & Zhao, L. (2004) Development of China's agricultural irrigation—problems and challenges, *Journal of Water Conservancy Economics*, 1, pp. 54–58 [in Chinese].

Hu, A. & Wang, Y. (2001) China's water allocation system reform: an observation of Dongyang-Yiwu water rights transaction, *China Water Resources (Beijing)*, 6, pp. 35–37 [in Chinese].

Hunan Provincial Government (2008) *Rules on Administrative Procedure of Hunan Province*. Available at http://www.chinacourt.org/flwk/show.php?file_id=129159 (accessed 29 July 2009) [in Chinese].

Lee, S. (2006) *China's Water Policy Challenges* (Nottingham: China Policy Institute, University of Nottingham). Available at http://www.nottingham.ac.uk/china-policy-institute/publications/documents/Discussion_Paper_13_LEE_China_Water_Policy_Challenges.pdf (accessed 10 April 2009).

Li, L. & Luo, H. (2006) Analysis of the gaming characteristics in China's water resources management, *China Population, Resources and Environment (Jinan)*, 2, pp. 37–41 [in Chinese].

Luo, J. (2008) On optimizing China's pollutant discharge permit system, *Academic Journal of Hehai University (Social Sciences Edn) (Nanjing)*, 3, pp. 32–36 [in Chinese].

Ministry of Commerce (2002) Conditions of Water Resources in China. Unpublished Report, 22 July 2002 [in Chinese].

Ministry of Construction and Ministry of Health (1996) *Rules on Hygiene Supervision of Domestic Potable Water*. Available at http://www.gov.cn/fwxx/bw/wsb/content_417703.htm (accessed 27 July 2009) [in Chinese].

Ministry of Environmental Protection (2001–2008) *Annual Report of Environment Status in China*, chapter on freshwater environment. Available at http://jcs.mep.gov.cn/hjzl/zkgb/ (accessed 14 November 2009) [in Chinese].

Ministry of Water Resources (2008a) *Provisional Rules on Water Quantity Allocation*. Available at http://www.mwr.gov.cn/tzgg/hdxw/200712281131465ff4bb.aspx (accessed 27 July 2009) [in Chinese].

Ministry of Water Resources (2008b) *Location Map of the Midline Part of South–North Water Transfer Project*, Available at http://nsbd.mwr.gov.cn/gcbj/zx/20080327102114e5ea54.aspx (accessed 28 July 2009) [in Chinese].

Ministry of Water Resources (2008c) *Regulations on Water-withdrawal Permit*. Available at http://www.mwr.gov.cn/zcfg/gz/20080417092300a52a9c.aspx (accessed 28 July 2009) [in Chinese].

Ministry of Water Resources (2008d) *Gazette of Water Resources in China 2007* (Beijing: Ministry of Water Resources). Available at: http://www.mwr.gov.cn/xygb/szygb/qgszygb/index.aspx (accessed 14 November 2009) [in Chinese]

Ministry of Water Resources (2001–2008) *China Water Resource Quality Annual Reports*. Available at http://www.mwr.gov.cn/xygb/szyzlnb/index.aspx (accessed on 11 April 2009) [in Chinese].

National Development and Planning Commission and Ministry of Constuction (1998) *Rules on Pricing Management of Urban Water Supply*. Available at http://xmxh.smexm.gov.cn/2005-6/2005619165308.htm, accessed 28 July 2009) [in Chinese].

National Statistics Bureau (2009) *China Statistical Year Book 2008* (Beijing: China Statistics Press) [in Chinese].

Standing Committee of National People's Congress (SCNPC) (1984) *Prevention and Control of Water Pollution Act*. Available at http://www.chinacourt.org/flwk/show.php?file_id=3543 (accessed 27 July 2009) [in Chinese].

Standing Committee of National People's Congress (SCNPC) (1988) *Water Act*. Available at http://www.chinawater.net.cn/law/WaterLaw.htm (accessed 27 July 2009) [in Chinese].

Standing Committee of National People's Congress (SCNPC) (1989a) *Communicable Disease Prevention Act.* Available at http://www.law-lib.com/law/law_view.asp?id=569 (accessed 27 July 2009) [in Chinese].

Standing Committee of National People's Congress (SCNPC) (1989b) *Environmental Protection Act.* Available at (Chinese and English versions) http://www.law-lib.com/law/law_view.asp?id=6229 (accessed 27 July 2009).

Standing Committee of National People's Congress (SCNPC) (1996) *Prevention and Control of Water Pollution Act (Amendment Act).* Available at http://www.chinawater.net.cn/law/wp-01.html (accessed 27 July 2009) [in Chinese].

Standing Committee of National People's Congress (SCNPC) (2002) *Water Act.* Available at http://www.chinawater.net.cn/CWSNews/newshtm/w020830-1.htm (accessed 27 July 2009) [in Chinese].

Standing Committee of National People's Congress (SCNPC) (2007) *Civil Procedure Act.* Available at http://www.gov.cn/flfg/2007-10/28/content_788498.htm (accessed 27 July 2009) [in Chinese].

Standing Committee of National People's Congress (SCNPC) (2008) *Prevention and Control of Water Pollution Act (Amendment Act).* Available at http://www.chinacourt.org/flwk/show.php?file_id=125455 (accessed 27 July 2009) [in Chinese].

State Council (1994) *Regulations on Municipal Water Supply.* Available at http://www.ynwater.com/http_low/national/3.htm (accessed 27 July 2009) [in Chinese].

State Council (2006) *Regulations on Water-withdrawal Permit & Collection of Water Resource Fee.* Available at http://www.gov.cn/zwgk/2006-03/06/content_220023.htm (accessed 28 July 2009) [in Chinese].

State Council (2007) *Regulations on Public Access to Government Information.* Available at http://www.gov.cn/zwgk/2007-04/24/content_592937.htm (accessed 28 July 2009) [in Chinese].

State Council (2009) *Comprehensive Work Plan on Energy Conservation and Pollutant Emission Reduction.* Available at http://news.xinhuanet.com/politics/2007-06/03/content_6191519.htm (accessed 13 April 2009) [in Chinese].

State Development and Reform Commission and Ministry of Water Resources (2003) *Rules on Pricing of Water Supply from Water Conservation Projects.* Available at http://www.bzwater.gov.cn/Article_Show.asp?ArticleID=47 (accessed 28 July 2009) [in Chinese].

State Environmental Protection Administration (2006) *Provisional Rules on Public Participation in Environmental Impact Assessment.* Available at http://www.mep.gov.cn/info/gw/huanfa/200602/t20060214_74178.htm (accessed 29 July 2009) [in Chinese].

State Planning Commission (2001) *The Administrative Instructions on Water Pricing Reform in the Agricultural Sector.* Available at http://www.jmsedu.net/info/jiaoyu/05/Product2/Law/22_Price/22_Price2431.htm (accessed 11 April 2009) [in Chinese].

US Department of Commerce, International Trade Administration (2005) *Water Supply and Wastewater Treatment Market in China* (Washington, DC: US Department of Commerce).

World Bank (2007) *Water Supply Pricing in China: Economic Efficiency, Environment and Social Affordability* (New York, NY: The World Bank).

World Bank (2008) *World Development Indicators Database 2007* (New York, NY: The World Bank). Available at http://web.worldbank.org/WBSITE/EXTERNAL/DATASTATISTICS/0,,contentMDK:21298138~pagePK:64133150~piPK:64133175~theSitePK:239419,00.html/.

The Water–Energy Puzzle in Central Asia: The Tajikistan Perspective

MURODBEK LALDJEBAEV

Lee Kuan Yew School of Public Policy, National University of Singapore, Singapore

ABSTRACT *The intricacy, interrelatedness, and complexity of the issues surrounding the management and use of Central Asia's natural resources call for a careful analysis of each issue. In this paper, the scope of the discussion will be restricted mainly to water–energy issues, and the focus will be primarily on Tajikistan. The significance of the discussion is intended for the ordinary people who suffer the consequences of the unresolved issues in the water and energy sectors. This paper aims to contribute towards demystifying the water–energy puzzle by searching for the sources of the problems as well as the avenues for their resolution.*

Introduction

The coming together on 28 April 2009 of the heads of the Central Asian countries—Kazakhstan, Kyrgyzstan, Tajikistan, Turkmenistan, and Uzbekistan—to discuss the issue of the Aral Sea 'ended with the signing of an agreement without any specific detail on transnational water management' (Marat, 2009). Albeit an important one, the issue of the Aral Sea is one among many in the region that remain unresolved. The intricacy, interrelatedness, and complexity of the issues surrounding the management and use of the region's natural resources call for a careful analysis of each issue. In this paper, the scope of the discussion will be restricted mainly to water–energy issues, and the focus will be primarily on Tajikistan. The rationale for such a concentration stems from the author's interest in the water–energy problem of Tajikistan. The significance of the discussion, however, is intended for ordinary people who suffer the consequences of the unresolved issues in the water and energy sectors. This paper aims to contribute towards demystifying the water–energy puzzle by searching for the sources of the problems as well as avenues for their resolution.

The paper is organized in five main sections. The first discusses the mechanism for the sharing of the natural resources of the Central Asian region as devised and implemented under the Soviet Union. The discussion of the failure of the Soviet mechanism and the resulting water–energy puzzle is then discussed. The subsequent two sections review the respective energy crises of 2008 and 2009; while the next objectively assesses the positions of Central Asian countries on the water–energy nexus and the approaches to its resolution. Finally, possible solutions to address the complex water–energy problems of the region are outlined.

Water for Irrigation and Electricity in Soviet Central Asia

The principal sources of the region's water are two major rivers—Syr Darya and Amu Darya—which originate in the mountains of Kyrgyzstan and Tajikistan, respectively, and cross the borders of these countries, flowing through the territories of Uzbekistan, Turkmenistan, and Kazakhstan and finally drain into the Aral Sea. For centuries, these rivers have served as a source of sustenance for the people who relied on the water supply for domestic use and agricultural activity.

The rich reserves of oil and natural gas, however, are found in the territories of Kazakhstan, Turkmenistan, and Uzbekistan. These energy sources have become crucial for the development and improvement of people's living conditions over the past two centuries.

The geographic distribution of the natural resources necessitates the interdependency of the countries in terms of sharing these resources. A plainly obvious mechanism for resource sharing in this context is for the upstream countries of Kyrgyzstan and Tajikistan to ensure a continuous flow of water to the downstream countries, and for the downstream countries of Kazakhstan, Turkmenistan, and Uzbekistan to channel fuel and gas to their upstream neighbours. This overall framework of resource sharing is simple and logical. Its realization in practice, however, is a challenge that is becoming increasingly complicated and difficult to resolve. Yet, it was handled without much difficulty when the countries were part of the former Soviet Union (USSR).

Under the former Soviet Union, the five Central Asian republics adhered to the policies dictated from Moscow. The water and energy policies were region-wide rather than republic-based, with priorities set at the country level (the USSR as a whole). The mechanism of natural resource sharing was implemented with reference to agricultural objectives, which primarily focused on the production of cotton and wheat. The emphasis on cotton production, in particular, emerged strongly in the 1960s as Moscow embarked on economic development via improved agricultural infrastructure. For this purpose

> a system of canals and pumps was constructed to withdraw water from the Amu Darya and Syr Darya before their discharge in the Aral Sea, and to convey the water to remote desert areas of Kazakhstan, Turkmenistan and Uzbekistan. (Dinar *et al.*, 2007, p. 288)

Similarly, the irrigated share of the southern lands of Tajikistan was expanded to allow for greater cultivation of cotton.

This increase in the area of irrigated land throughout the region raised the need to maintain a continuous supply of water during the irrigation season. A series of reservoirs were then built in the upstream reach of Amu Darya and Syr Darya to collect water for the purpose. Of particular significance in terms of capacity are Toktogul ($19.5 \, \text{km}^3$) in Kyrgyzstan, and Nurek ($10.5 \, \text{km}^3$) and Kayrakum ($4.16 \, \text{km}^3$) in Tajikistan (Severskiy *et al.*, 2005). Another large reservoir, Chadarya ($5.7 \, \text{km}^3$), is located in the downstream reach of Syr Darya in the territory of Kazakhstan.

In addition to water accumulation, these giant reservoirs performed an important economic function of generating electricity. The installed capacity of the Toktogul hydropower station (HPS) was 1200 megawatts (Zozulinsky, 2007). Nurek HPS has a total installed capacity of 3000 megawatts, the largest in the region (Asian Development Bank (ADB), 2008). The capacity of Kayrakum is relatively smaller at 126 megawatts (State Agency for Hydrometeorolgy of Committee for Environmental Protection under the Government of the Republic of Tajikistan, 2008). To put the numbers into perspective, Toktogul HPS accounts for 93% of electricity production in Kyrgyzstan (United Nations Development Programme (UNDP), 2009), while Nurek HPS generates over 70% of the electricity in Tajikistan (ADB, 2008).

Given these large scales both in terms of water collection and electricity generation capacity, it is not difficult to understand the impact of the reservoirs and HPSs on the economic development of the region. Economic gains in turn spurred a higher standard of living for the people in Central Asia. Yet, these grand-scale projects were not devoid of social and environmental costs. For one thing at least, the Aral Sea stands as a live witness to the negative impact of economic development without much regard to the environment. These implications, however, will be dealt with in more detail below.

Water and electricity infrastructure was operationalized to service the economies of all five Central Asian republics. The upstream Kyrgyzstan and Tajikistan collected water during the non-irrigation period (essentially in winter) and released it downstream to Kazakhstan, Turkmenistan, and Uzbekistan during the irrigation period. Water allocation among the republics was mainly regulated from Moscow. As the collection of water meant reduced electricity generation for the upstream republics, they were 'compensated by stable supplies of mineral energy resources' from the downstream republics (Mamatkanov, 2008, p. 141). The release of water in summer generated surplus of electricity, which was transmitted to power the irrigation pumps downstream. In winter, the upstream republics received electricity from their downstream neighbours. 'Moscow covered the costs of operating and maintaining the dams, reservoirs, canals and irrigation pumps' (International Crisis Group (ICG), 2002, p. 7). It should be noted that

> during the Soviet period, Kyrgyzstan, Uzbekistan, Tajikistan, Turkmenistan and the five southern provinces of Kazakhstan (the northern provinces of Kazakhstan were part of the Russian energy grid) were all part of the United Central Asian Energy System, which was managed by the USSR Ministry of Energy. (ICG, 2002, p. 7)

This mutually beneficial mechanism of resource sharing facilitated hand-in-hand growth of the Central Asian region during the Soviet era. With the dissolution of the USSR, however, central planning and control faded away, and with it the regional projects and mechanisms. Suddenly, the level of uncertainty became elevated in every aspect of the social, economic, and political landscape of the newly independent states. It is remarkable to observe that one of the first things the countries did in this uncertain environment was to agree on water allocation levels. The countries agreed for the immediate period to maintain the existing mechanism inherited from Soviet times. The settlement, known as the Almaty Agreement of 1992, was expected to hold 'until the states could reach a solution amenable to all parties' (Dinar et al., 2007, p. 294).

In retrospect, nonetheless, the countries have experienced many challenges to reconcile their differences and come to mutually favourable terms in sharing the region's natural resources. In the following sections, some of those major challenges will be discussed with the example of Tajikistan.

The Water–Energy Puzzle: Tajikistan and its Neighbours

The water and energy inter-linkages in the context of Tajikistan are exhibited in the use of water as the main source of electricity production, on the one hand, and the use of electricity to pump water for irrigation and domestic use, on the other hand. The use of water in the production of other forms of energy, such as thermal and nuclear energy, is, however, minimal because the thermal power plants are practically on the verge of complete disuse, whereas nuclear technologies for civilian purposes have not yet been developed in the country. Given these interrelationships, the water and energy link appears straightforward at first glance, but in fact, it is not that simple. The complexity of the problem emerges at the regional level, where the countries of Central Asia depend on each other for their natural resource needs. It is at this level from which the water–energy puzzle originates.

Following their independence from the Soviet Union, the five countries of Central Asia followed their path to development in a less integrated way than before. The once centrally designed plans for regional development had to be revised as each state identified its own national priorities. The transition process was especially painful for Tajikistan, because of the civil war that broke out almost immediately after it gained its independence in the early 1990s and lasted for almost a decade.

Despite the continued instability in the country, Tajikistan maintained its presence and participation in regional cooperation. In the water sector, the signing of the Almaty Agreement in 1992 and the establishment of the Interstate Commission for Water Cooperation (ICWC) were significant points of departure on the new road towards management of the region's water resources. The Almaty Agreement fixed the water allocations at the existing levels determined under the former Soviet Union, and the ICWC was mandated to manage, monitor, and report on the status of those allocations. In the following years, environmental, social, and economic issues gained prominence in the Aral Sea Basin, leading to the signing of agreements among the riparian states to deal with these issues. As Dinar *et al.* (2007, p. 295) account:

> Additional organizations were created between 1993 and 1995 to support the management of the Aral Sea Basin. These included the Interstate Council on the Aral Sea Basin (ICAS) that was formed to develop policies and proposals for the management of the Aral Seas Basin (Peachy, 2004); the International Fund for the Aral Sea (IFAS), designed to manage contributions and to finance program activities (Mukhammadiev, 2001); the Sustainable Development Commission (SDC) formed to ensure that socio-economic issues were considered by ICAS when determining new policy, and the Executive Committee of the ICAS (EC-ICAS), which was given the responsibility of implementing programs set forth by the Aral Sea Basin Program (ASBP). ... The ASBP, initiated in 1994, is a consortium of international organizations such as UNDP, UNEP, the World Bank, and the EU [European Union]. It is aimed at identifying long-term solutions for the basin's wide-ranging

problems (environment, water management, rehabilitation of the disaster zone around the lake).

In addition to the above, numerous other agreements were concluded among the riparian countries within the same period. As Dinar *et al.* (2007) refer to the estimate by Peachy (2004), between 1991 and 1994 the number is over 300 informal agreements concerning the Aral Sea Basin. Other settlements and deals followed in subsequent years. However, as the countries progressed in their national consolidation efforts and attained some level of economic development, their approach to regional cooperation on water resources gradually shifted to that of maximizing individual gains from the agreements. As a result, no conclusive agreement was reached by any of the riparian states to manage the region's water resources in an efficient and equitable manner.

Instead, what has developed over the past decade is an intractable puzzle that dominates the landscape of Central Asia's natural resource management. The Soviet mechanism of trading collected water and electricity surplus of the upstream countries in the summer with the energy supplies of the downstream countries in winter has proven to be unreliable. Meeting their internal demand for electricity and gas, the downstream countries could not ensure a reliable supply for the upstream countries. They were also pressed to increase the charges in order to pursue their economic development objectives. The upstream countries, also facing development challenges, began to use more water for electricity generation in order to reduce the costs (by lesser imports), meet local demand and, in summer, increase their share of exports.

The countries got entangled in disputes over the allocation of water use, particularly in relation to the time and purpose of such use. As is eloquently expressed by the Uzbekistan Embassy in Singapore (2008, p. 1):

> increase of water discharge in wintertime leads to flooding of useful territories, demolition of houses and creation of other extraordinary situations and damages which account for millions of dollars. The work of water reservoirs during the summer season in the water accumulation mode creates a shortage of water resources for agricultural production, reduction of the sown area for crops and agricultural output, and consequently, deterioration of living conditions of population.

The increasing release of water is from Toktogul in the territory of Kyrgyzstan. In its turn, 'Bishkek also accused Tashkent of effectively causing the flooding as it does not stick to its part of the barter agreement—either by providing less gas than agreed, or by cutting supplies altogether' (ICG, 2002, p. 14).

In addition to the conflicting timing of water use, the general use of water for irrigation is also the subject of heated disagreement. The blame game that rests on excessive use of water beyond the specified quotas is a closed chain that binds all five states. 'Turkmenistan is using too much water to the detriment of Uzbekistan, which in turn has been accused by Kazakhstan of taking more than its share. Kyrgyzstan and Tajikistan say that the three downstream countries are all exceeding quotas' (ICG, 2002, p. ii).

In this water–energy dilemma, all five countries are incurring costs in some way or another. While the lack of cooperation is hurting everyone, it appears that Tajikistan is most vulnerable of all. Its relative disadvantage, as compared with Kyrgyzstan, is that Tajikistan cannot control the flow of Amu Darya to a similar extent as Kyrgyzstan controls the flow of Syr Darya. Before all, almost half of the water flow in Amu Darya is due to its Panj tributary, on which there are no reservoirs. As for the second half, the Vakhsh River bears the Nurek reservoir. However, Nurek HPS must produce an electricity surplus in the summer in order to gain much-needed export money. Also, the thirsty cotton fields of southern Tajikistan depend on the release of water from Nurek. Finally, Nurek cannot store a significant flow of Amu Darya as it threatens to wash away the dam.

Given this state of affairs, Tajikistan lacks an enforceable mechanism to put pressure on its downstream neighbours (particularly, Uzbekistan) to comply with water, gas, and electricity agreements. Hence, in the water–energy game Tajikistan emerges as the weakest of all.

In order to strengthen its position, Tajikistan's growth strategy is to realize its hydropower potential. As Severskiy *et al.* (2005) state, 'of all the countries of the Aral Sea Basin Tajikistan has the greatest hydroelectric potential of all the countries in the Aral Sea Basin—more than 52 000 GWh/year'. In the grand plans of the country, it is the construction of a series of hydropower plants that are envisaged to rescue the country from energy dependency on its oil- and gas-rich neighbours (Parshin, 2003). Moreover, given the expected full-scale realization of energy generation and thereby increased exports of electricity, the surplus will be looked at to contribute significantly towards economic growth and, ultimately, raising people's standard of living.

The construction of hydropower plants, a plan also pursued by neighbouring Kyrgyzstan, however, is not acceptable to the downstream countries. The primary concern is the expected shortage of water, which will result from water accumulation in the upstream reservoirs. This will increase the downstream countries' dependence on water, and in response they 'have outlined plans to build their own reservoirs, further complicating the development of a coherent regional system of management' (ICG, 2002, p. 2).

Another important concern raised by Uzbekistan in relation to the planned constructions has been that of environmental impact. In his addresses to his upstream neighbours as well as to the international community, President Islam Karimov stressed the importance of conducting a thorough environmental feasibility study of the large projects. Furthermore, he stated that these projects should be approved by all riparian countries. Addressing the Uzbek government meeting on 13 February 2009, Karimov noted in particular that if the results of the international and independent evaluations turn out to be positive, then Uzbekistan can take part in the investment in those projects (Hamidova, 2009).

This last change in attitude suggests that there is good scope for cooperation among the Central Asian countries to work and grow in partnership rather than inflict suffering on each other. However, as the next section will discuss, there is still a long way before the hardships can be mitigated.

Winter Energy Crisis in Tajikistan: 2008 Revisited

The winter of 2008 was characterized by exceptionally cold weather and heavy snowfall. Extremely low temperatures throughout the country led to a surge in the demand for heating. The need was especially acute in the capital city of Dushanbe, which was unprepared for the harsh winter. Central heating systems that were few in number went out of service and, thus, electricity served as the primary source of heating. With the breakdown of the electricity infrastructure, due its rundown conditions, the city sank into darkness and cold.

The level of water in the reservoirs dropped significantly as a result of the increased outlet of water to produce additional kilowatts. However, the amount of water was far from enough. As no substantial improvement of the situation was observed, households began to adopt co-opting strategies. The use of gas increased as it was required for both cooking and heating. Thus, a shortage of electricity triggered a surge in the use of gas. But gas was also in short supply. Therefore, as the supply of both electricity and gas was interrupted, sometimes for many hours, households had to search for wood, coal, paper boxes, and other materials for outdoor fires to cook their food and to warm themselves for a while (Panfilova, 2008). Conditions were so harsh that foreign embassies closed and sent their diplomats back to their respective countries. Unfortunately, there was no escape for the local people, who had to endure the hardship and struggle for their own survival.

All these developments ultimately led to grave consequences. As the power cuts continued throughout the city, offices were also closed. Exceptions were facilities of social significance, which had to function. Hospitals were particularly of importance to keep operational. These structures were the priority for electricity provision. Nonetheless, they had also suffered from the extreme lack of electricity for heating, including performing medical procedures (especially surgical operations). The great misfortune was that newborns died in the maternity hospitals. The reports of different agencies state different figures, but the number of infants who died during the winter energy crisis is estimated to be above two hundred (CentrAsia, 2008).

The supply of water to households, hospitals, schools and other civil entities sharply decreased. This resulted in the extreme cold bursting pipes and causing the collapse of the pumping stations.

Clearly, the consequences of the energy crisis alerted the authorities to the emergent humanitarian crisis. To cushion the hardship and prevent a humanitarian catastrophe, the government appealed to the international community for urgent assistance. The response came in the form of food and goods of immediate need. The cost of the crisis, however, was estimated to be around US$250 million (Isamova, 2008). Many countries and organizations then pledged their support to the country, but this is yet to be confirmed and disbursed (Office for the Coordination of Humanitarian Affairs (OCHA), 2009).

Another Energy Crisis: 2009 and Counting

The winter of 2009 marked another severe energy crisis in Tajikistan, despite the many assurances by the authorities that the last year's crisis was not to be repeated. This crisis was very similar to the previous year's scenario. Low temperatures and the shortage of water in rivers and reservoirs led to significant electricity shortages and the eventual need for rationing. The difference from last year, however, was that temperatures were at or above historical levels (UNDP, 2009). Given that, the likelihood of the crisis should have been less. However, it was more due to other factors that people in Tajikistan had to suffer another harsh winter. These factors, explored in detail below, were in fact also contributors to the 2008 energy crisis. However, they are discussed here because the effect of the main factor—low temperatures and heavy snowfall—can be assumed to be relatively minimal in the 2009 crisis.

The events that led in some way or another to the emergence of the 2009 energy crisis in Tajikistan can be attributed to political and economic factors. These factors are also intertwined and influence one another. Yet, it is the interstate relations with neighbouring

Uzbekistan that effectively led to the start of the crisis. Following the chain of events that began to unfold from December 2008 and continued over the winter months, the root causes of the crisis become evident.

Just before the new year, Uzbekistan more than doubled the price of natural gas exported to Tajikistan. After meetings of representatives of Tajikgas and Uztransgas, the price was renegotiated and decreased from US$300 to US$240 per $1000\,m^3$. This change was an increase of US$90 from the previous year (Naumova, 2009a). The price hike was necessitated by the spread and deepening of the global financial crisis.

In January 2009 the situation of electricity provision in Tajikistan suddenly worsened as Uzbekistan discontinued the transmission of electricity from Turkmenistan through its territory. This was done on the grounds that the agreement on the terms of transit was not extended into the new year (Daily New Bulletin, Moscow, 2009). A number of meetings occurred between Tajikistan and Uzbekistan before transmission was restored. However, the process took over a month-and-a-half, as a result of which the country slid into the energy crisis. Strict electricity rationing was enforced for rural areas at 2 hours a day, while in the capital city blackouts stretched to 9 hours a day (Hamroboyeva & Hasanova, 2009).

The supply of natural gas was also cut by Uzbekistan, which exacerbated the situation. It was explained that the reduction in gas supply was due to non-payment of the outstanding debts by Tajikistan (EurasiaNet, 2009).

Such a chain of events evolves each year and the population remains vulnerable every winter. While the events appear to be technical and easy to resolve, a broader perspective reveals the true complexity of the issues. As was alluded to above, the bone of contention is the water–energy puzzle that must be resolved at a regional level. Unless, and until, this puzzle is unravelled, the people of Tajikistan will continue to suffer from a shortage of energy supplies during the winter months. As discussed below, however, cooperation at the regional level has not yet found the direction that could lead to the resolution of the water–energy challenge.

Regional Cooperation: A Continuing Polemic

On 12 June 2009 a regional roundtable was organized in Dushanbe by the Institute for War and Peace Reporting (IWPR), the participants of which were experts from Central Asian countries, representatives of their respective governments, as well as international organizations and the mass media. The gathering set out to seek for an answer to the currently most disputable question: 'Water and energy problems in Central Asia: is compromise possible?' As the Executive Manager of IWPR in Central Asia, Mr Abakhon Sultonnazarov told the Asia-Plus correspondent: 'recently a real war is being waged in [the] mass media that concerns these problems, and therefore, it is necessary to review the objectivity and transparency of the coverage of water–energy problems in [the] mass media' (Hasanova, 2009).

The information war, which has recurred frequently in the mass media, is a tangible sign of the lack of cooperation among Central Asian countries to resolve the water–energy issues. The blame game has become the routine rhetoric in news reporting and online articles. Aside from drawing the attention of the wider audience to the problems, the mass media is, however, unlikely to achieve anything more. The vacuum created by the lack of constructive dialogue between the leadership of the respective countries cannot be filled with continuing polemics that lead to more disputes rather than to an agreement. And for the agreement to be settled, all stakeholders from every country should sit at the negotiating table and find a mutually satisfying compromise.

It should not be concluded that no meetings have been conducted towards this purpose. On the contrary, many discussions have taken place to tackle the existing problems in the water–energy sectors at the regional level. Apart from piecemeal deals, however, nothing substantial has been agreed upon or implemented. As the most recent meeting of the heads of Central Asian countries in Almaty on 28 April 2009 demonstrated, 'none of the Central Asian [leaders are] yet prepared for compromises, and that a solution to the dispute therefore remains a distant prospect' (Institute for War and Peace Reporting (IWRP), 2009).

It is becoming more evident that a regional solution is bound to be a difficult one. Statements by the Presidents of Uzbekistan and Tajikistan over the past few months of 2009 have evolved from seeking external involvement in the resolution of the disputes to opting for bilateral negotiations. In his address to the participants of the regional political dialogue between Troyka EC and Central Asia at the level of the Ministries of Foreign Affairs, President of Tajikistan Emomali Rahmon stated that

> we believe that the solution to this problem can be achieved through the constructive dialogue at the negotiations table. I am personally for reaching the consensus without intermediaries and we are ready for an open dialogue. I believe, we can come to an agreement. (Naumova, 2009b)

The position of Kyrgyzstan in this respect is to enter into equal negotiations with all its neighbouring countries. The Prime Minister of Kyrgyzstan, Igor Chudinov, stated that 'the problem of water and energy can be resolved not by [the] publishing of articles on the newspapers and internet sites but at the negotiations table' (Farghony, 2009).

Kazakhstan recognizes the intensity of the dispute, but does not take sides. At the roundtable in Dushanbe on 12 June 2009, Bulat Ayelbaev, an expert from Kazakhstan, remarked that the

> establishment in Central Asia of a water–energy consortium will help to resolve important regional problems including agricultural, hydroenergy as well as heating– electricity complexes. Kambarata HPS of Kyrgyzstan, in the construction of which Kazakhstan is prepared to invest, could become a basis for such a consortium. (Chorshanbiev, 2009b)

As these statements signify, Kyrgyzstan and Kazakhstan take a regional approach to resolving the water–energy puzzle, whereas the positions of Tajikistan and Uzbekistan reveal a tendency towards bilateralism. The position of Turkmenistan, which used to be that of non-involvement, has also changed. Though its approach and involvement in the resolution of the dispute is less straightforward, President of Turkmenistan, Gurbanguly Berdymuhammedov, insisted in the Almaty summit on 12 April 2009 that 'no measures to change the water and energy balance must be taken unless they have been agreed' (IWRP, 2009).

In the overall scheme, however, it appears that all the Central Asian countries recognize the need for negotiations to determine the mechanism for mutually beneficial sharing of the region's water and energy resources. In practice, however, things have hardly moved off the ground. What remains the reality today is a sort of continuing polemic that runs back and forth between countries, and this rhetoric in turn is galvanized in the mass media.

Yet another roundtable concluded with the resolutions to be made public on 15 June 2009. Of the five Central Asian countries, only three were represented; Turkmenistan and Uzbekistan did not send their representatives for unknown reasons. As the discussions proceeded, Suhrob Sharipov, an expert from Tajikistan, pointed out that 'if energy problems of Tajikistan and Kyrgyzstan are not resolved, they can lead to socio-political explosions. However, it remains unintelligible how the water–energy problem of the region will be addressed' (Chorshanbiev, 2009a).

Negotiations are without doubt a way of resolving disputes. However, there are other measures as well. In the next section, an attempt will be made to point out the possible avenues that could lead to the resolution of the long-time disputes from the perspective of Tajikistan.

Avenues to Solve the Water–Energy Puzzle

In search of solutions to its energy shortages, Tajikistan has considered two main options, which are not necessarily mutually exclusive. Both options rest on the long-term strategy of the country that aims to maximize the use of its hydropower potential. These two options essentially entail engineering solutions. The difference is in the scale of the projects. However, a third option in addition to the existing ones is suggested here by the author. It could make a notable contribution to the efficient use of energy resources. It should be noted, however, that such options as the diversification of energy sources (wind, biofuel, etc.), institutional reforms of the sectors, greater cooperation with neighbouring countries, the involvement of third parties, e.g. the European Union and/or Russia in the negotiations, and many others have already been proposed, and some of them are being implemented. The focus here is rather on the options that the government of Tajikistan identifies as the most strategically significant.

Option I. Construction of Roghun HPS

The first option is very costly and controversial. However, it is believed not only to satisfy the current and future needs of the country's electricity, but also to generate substantial surplus which could be exported for profit. The technical and financial feasibility studies were conducted at the time of the inception of the Roghun HPS during the time of the Soviet Union. However, a reassessment of such a feasibility as well as environmental and social impact analysis is required to identify and weigh the potential benefits, which may or may not then justify otherwise prohibitive costs. This extremely high level of uncertainty is also the cause of controversy with downstream countries. Neither side has presented any empirical evidence to justify its standpoint.

The Presidents of Tajikistan and Uzbekistan appealed to The World Bank Group to provide an independent assessment of the project. In his letter to the President of Uzbekistan on 23 April 2009, the President of The World Bank, Robert Zoellick, explained that the bank was responsible only for the preliminary evaluation of the feasibility of Roghun HPS. He further assured that every important aspect of the project would be carefully assessed by independent consultants. And the ultimate decision of whether or not such grand projects should be pursued would depend on the condition that all involved parties had no hesitations about the project (Zoellick, 2009).

Given the high levels of uncertainty and controversy, the realization of the construction of Roghun HSP will remain unknown until the evaluations are conducted. Nonetheless, Tajikistan has initiated the construction process using its own budget allocations. Such a decision, though early and unfounded, has in fact proven to be strategic as it triggered a reaction from the downstream countries and thus attracted international attention.

The short-term recommendation for this option, therefore, is to maintain an emphasis on the construction of Roghun HPS. In the long-term, however, Tajikistan should not place high stakes on the construction and in fact it should be prepared to abandon the project in case the social, environmental, and financial costs outweigh the potential benefits.

Option II. Construction of Small-scale HPSs

In contrast, the second option is neither too costly nor too controversial. Nonetheless, small-scale HPSs still should undergo rigorous scrutiny to identify and weigh the social costs and benefits. The recommendation for this option is that such projects must be identified and a systemic approach to their construction adopted. The systemic approach is important especially in view of the long-term. The small-scale HPSs must not be considered as short-term get-by solutions because, once constructed, the cost of undoing them would be very high and such an approach would be very inefficient. The long-term consideration will be to address, for example, how these small-scale HPSs can connect to the main grid and what share of the expected demand they will satisfy, etc.

Option III. Improve Efficiencies

In addition, a third option is proposed that is well known to professionals as well as the leadership of Tajikistan, but to the author's observation this option has received only 'token' attention.

The system, management, operation and maintenance, and other inefficiencies in water and electricity are frequently blamed upon the decaying infrastructure, a lack of incentives, staggering corruption, and other factors that inhibit the maximum use of the existing capacities in both sectors. Stark evidence of such improvements in the efficiencies have been the privatization of a small-scale HPS in one of the four provinces of Tajikistan, which in contrast to other parts of the country has ensured a higher quality of service. In the harsh winters of the past two years the residents of Gorno-Badakhshan Autonomous Oblast, who live in predominantly rural areas, experienced only mild interruptions in their supply of electricity.

What this case suggests is that improvements in the efficiencies are possible and can be done. But what it does not yet explain is what the impact has been on the welfare of the poor and/or otherwise disadvantaged people. Therefore, the recommendation for this option is to begin with an impact assessment of that exceptional case and to learn from its experience of improving the efficiencies of the existing electricity and water supply structures. Also of significance is the learning from the experiences of other countries that succeeded in this path.

This option is recommended as a systems approach that cuts across all sectors of the economy. By way of illustration, the efficiency improvements in the water and electricity sectors positively impact the agricultural sector. For example, by ensuring a fast and reliable water supply, the cost of waiting time will be reduced, thus contributing to overall profits. On the other hand, efficiency improvements in the agricultural infrastructure will lead to a reduced demand on water and electricity, thus contributing to water and energy conservation. The multiplier effect of the efficiencies across all sectors will result in better management of the resources and, by extension, an improvement in the life of the people.

Conclusions

Water remains a valuable strategic resource for Tajikistan. Though a complete readjustment is not an objective, the use of water is gradually shifting priority from irrigation to the production of electricity. As the country grapples with meeting its peak demands for energy in the cold months, the water–energy puzzle emerges as a critical challenge. In the complex set of interactions with other Central Asian countries, Tajikistan struggles to benefit from the equitable sharing of the region's natural resources. Though there are prospects for negotiation and agreement, there is a long way that lies ahead for this water-rich, energy-poor country.

To solve its energy problems and secure a reliable path of economic growth, Tajikistan seeks to realize its massive potential for hydropower generation. The construction of large and small hydropower plants is envisioned and/or planned. It is estimated that the surplus of electricity will be sold to other countries and the profits earned thereby would be reinvested in the country's economy in order to catalyse growth and, thus, improve the people's quality of life.

As these options stand at present, a considerable level of uncertainty is involved in their realization. These uncertainties, which are primarily to do with the expected social benefits and costs, engender substantial controversy on the part of neighbours as well as the international community. Impact assessments have already been proposed to minimize the uncertainties and facilitate informed decision-making.

Concurrently, it is proposed to emphasize the efficiency improvements of the existing capabilities as these have been evidenced to make a sizeable impact. The systems approach should be adopted to improve efficiencies across all sectors of the economy as their multiplier effect will result in larger improvements in people's standard of living.

References

Asian Development Bank (ADB) (2008) *Proposed Asian Development Fund Grant, Republic of Tajikistan: Nurek 500 kV Switchyard Reconstruction Project. Asian Development Bank.* October (Manila: Asian Development Bank). Available at http://www.adb.org/Documents/RRPs/TAJ/42189-TAJ-RRP.pdf (accessed 3 May 2009).

CentrAsia (2008) *Zamerzli v roddomakh* [Frozen in maternity hospitals], *CentrAsia*, 2 February. Available at http://www.centrasia.ru/newsA.php?st=1202546820 (accessed 29 April 2009).

Chorshanbiev, P. (2009a) *Dostizhenie energeticheskoy nezavisimosti dlya Tajikistana prevratilos v natsionalnuyu ideyu, a kto protiv—tot predatel* [Attainment of energy independence for Tajikistan turned into a national idea, but whoever is against Roghun is a traitor], *Asia-Plus*, 12 June. Available at http://www2.asiaplus.tj/news/19/52976.html (accessed 12 June 2009).

Chorshanbiev, P. (2009b) *Kazakhstan ne podderzhit Uzbekistan v reshenii vodno-energeticheskoy problemy Centralnoy Azii* [Kazakhstan will not back Uzbekistan in resolving water–energy problems in Central Asia], *Asia-Plus*, 12 June. Available at http://www2.asiaplus.tj/news/19/52992.html (accessed 12 June 2009).

Daily New Bulletin, Moscow (2009) Turkmenistan suspends electricity export to Tajikistan, *iStockAnalyst*, 4 January. Available at http://www.asiaplus.tj/news/31/44747.html (accessed 3 May 2009).

Dinar, A., Dinar, S., McCaffrey, S. & McKinney, D. (2007) *Bridges Over Water: Understanding Transboundary Water Conflict, Negotiation and Cooperation* (Singapore: World Scientific).

EurasiaNet (2009) Uzbekistan: Tashkent cuts gas to Tajikistan, again, *Eurasianet.org*. 3 March. Available at http://www.eurasianet.org/departments/news/articles/eav040309b.shtml (accessed 3 May 2009).

Farghony, M. (2009) *Mavqeyi Bishkek dar bahsi obi Osiyoi Miyona* [Bishkek's position at the water dispute in Central Asia], *Radio Liberty*, 6 July. Available at http://www.ozodi.org/content/article/1750415.html (accessed 6 July 2009).

Hamidova, P. (2009) *Budem druzhit stramani?* [Let's make country friendship?], *Asia-Plus*, 19 February. Available at http://www.asiaplus.tj/articles/48/3082.html (accessed 20 February 2009).

Hamroboyeva, N. & Hasanova, M. (2009) Electricity rationing introduced in Tajik capital, *Asia-Plus*, 27 January. Available at http://www.asiaplus.tj/en/news/31/46010.html (accessed 3 May 2009).

Hasanova, M. (2009) CA experts gather in Dushanbe tomorrow to discuss water and energy problems, *Asia-Plus*, 11 July. Available at http://www.asiaplus.tj/en/news/50/52896.html (accessed 11 July 2009).

Institute for War and Peace Reporting (IWRP) (2009) *Water Dispute Unresolved at Central Asian Summit.* 1 May (IWRP). Available at http://www.iwpr.net/?p=bca&s=b&o=352205&apc_state=henh (accessed 14 July 2009).

International Crisis Group (ICG) (2002) *Central Asia: Water and Conflict*, Asia Report No. 34. 30 May (Osh/Brussels/ICG). Available at http://www.crisisgroup.org/library/documents/report_archive/A400668_3005 2002.pdf (accessed 3 May 2009).

Isamova, L. (2008) *Tajikistan prosit ekstrennoy pomoshi v svyazi s cilnymi morozami* [Tajikistan appeals for emergency assistance due to severe cold], *RIA News*, 6 February. Available at http://www.rian.ru/elements/20080206/98489021.html (accessed 3 May 2009).

Mamatkanov, D. M. (2008) Mechanisms for improvement of transboundary water resources management in Central Asia, in: J. K. Moerlins, M. K. Khankhasayev, S. F. Leitman & E. J. Makhmudov (Eds) *Transboundary Water Resources: A Foundation for Regional Stability in Central Asia*, pp. 141–152 (Dordrecht: Springer).

Marat, E. (2009) 'Water Summit' in Central Asia ends in stalemate, *The Jamestown Foundation*, 30 April. Available at http://www.jamestown.org/programs/edm/single/?tx_ttnews%5Btt_news%5D=34931& tx_ttnews%5BbackPid%5D=27&cHash=769e974d35 (accessed 1 May 2009).

Mukhammadiev, B. R. (2001) Legal aspects of interstate cooperation for transboundary water resources management in the Aral Sea basin, Proceedings of AWRA/IWLRI–University of Dundee Inernational specialty conference, 6–8 August. Available at: http://www.awra.org/proceedings/dundee01/Documents/MukhammadievPoster.pdf

Naumova, V. (2009a) *V Tajikistane podnimayutsya ceny na prirodny gas* [The prices for natural gas are rising in Tajikistan], *Asia-Plus*, 1 January. Available at http://www.asiaplus.tj/news/31/44747.html (accessed 3 May 2009).

Naumova, V. (2009b) *Prezident Tajikistana: Ni odin project Tajikistana v gidroenergetike ne budet napravlen protiv nashikh sosedey* [President of Tajikistan: no hydroenergy project in Tajikistan will be directed against our neighbors], *Asia-Plus*, 30 May. Available at http://www.asiaplus.tj/news/19/52321.html (accessed 30 May 2009).

Office for the Coordination of Humanitarian Affairs (OCHA) (2009) *Tajikistan—Compound Crisis—January 2008.* 4 May (OCHA). Available at http://ocha.unog.ch/fts/reports/daily/ocha_R10_E15489_asof___0905030206.pdf (accessed 4 May 2009).

Panfilova, V. (2008) *Silnye Morozy i Energetichesky Krizis* [Sever cold and energy crisis], *Arba.ru*, 16 January. Available at http://www.arba.ru/news/3736 (accessed 3 May 2009).

Parshin, K. (2003) Tajik power plans still tread water, *EURASIANET*, 15 December. Available at http://www.eurasianet.org/departments/business/articles/eav121503.shtml (accessed April 30, 2009).

Peachy, E. J. (2004) The Aral Sea basin crisis and sustainable water resource management in Central Asia, *Journal of Public and International Affairs*, 15, 1–20.

Severskiy, I., Chervanyov, I., Ponomarenko, Y., Novikova, N. M., Miagkov, S. V., Rautalahti, E. & Daler, D. (2005) *Aral Sea—GIWA Regional Assessment 24* (Kalmar: University of Kalmar on behalf of UNEP).

State Agency for Hydrometeorolgy of Committee for Environmental Protection under the Government of the Republic of Tajikistan (2008) *Second National Communication of the Republic of Tajikistan under the United Nations Framework Convention on Climate Change* (Dushanbe: United Nations Framework

Convention on Climate Change (UNFCCC)). Available at http://unfccc.int/resource/docs/natc/tainc2.pdf (accessed 3 May 2009).

United Nations Development Programme (UNDP) (2009) *Central Asia Regional Risk Assessment: Responding to Water, Energy, and Food Insecurity*. January (UNDP Tajikistan). Available at http://www.undp.tj/files/CARRA_2009_eng.pdf (accessed 3 May 2009).

United Nations Development Programme (UNDP) *et al.* (2005) *Aral Sea, GIWA Regional Assessment 24* (Kalmar: University of Kalmar).

Uzbekistan Embassy in Singapore (2008) *Problems of Aral Sea and Water Resources*. 28 August (Singapore: Embassy of the Republic of Uzbekistan). Available at http://www.uzbekistan.org.sg/aral-sea/ (accessed 3 May 2009).

Zoellick, R. (2009) *Poslanie na imya prezidenta Uzbekistana Islama Karimova* [Letter to President of Uzbekistan, Mr. Islam Karimov], *Press Service of the President of the Republic of Uzbekistan*, 23 April. Available at http://www.president.uz/#ru/news/show/press/poslanie_na_imya_prezidenta_uzbekistana_/hightext/роберт_зеллик/ (accessed 24 April 2009).

Zozulinsky, A. (2007) *Kyrgyzstan: Power Generation and Transmission*. March (Bishkek: US Embassy). Available at http://bishkek.usembassy.gov/uploads/images/rXcSjKDyhKkWIbSpUUIegg/KG_07_Power_Generation_Report.pdf (accessed 3 May 2009).

The Emergence of Water as a 'Human Right' on the World Stage: Challenges and Opportunities

ARJUN KUMAR KHADKA

Lee Kuan Yew School of Public Policy, National University of Singapore, Singapore

ABSTRACT *Since the end of the Cold War, the world has moved towards democratization, globalization, liberalization and privatization in an enthusiastic and complex fashion. Such an environment could be beneficial for the promotion and protection of human rights at regional, national and international levels. In practice, human rights are basic things, such as the right to food, the right to a home and the right to freedom. However, a right to water is not mentioned as a human right in the various international declarations, such as the Universal Declaration of Human Rights. In this regard, the Committee on Economic, Social and Cultural Rights, which was established to oversee the implementation of the Covenant on Economic, Social and Cultural Rights, presented a document at the UN 29th Session, in Geneva, Switzerland, in 2002. The committee re-interpreted Articles 11 and 12 of the International Covenant on Economic, Social and Cultural Rights, and concluded that 'water' can be considered to be a 'human right'. After that conclusion, water is legally emerging as a human right in many countries. However, there are significant challenges and opportunities for implementing the idea that water is a human right.*

Introduction

Background to the Study

Human rights are basic fundamental rights concerning issues such as life, liberty, food, shelter, and health. These are equal rights for both men and women, which encourage social progress and better standards of living and also the foundation of freedom, justice, and peace. Human rights recognize the inherent dignity and inalienable rights of all members of the human family (United Nations, 1948). They are universal, inalienable, and essential for everyone without limitations and restrictions, irrespective of where in the world one lives.

Human rights are a symbol of justice for survival and progress because if people lack human rights, it creates a disparate and unjust society. At the end of the 20th century, they have become a core agenda in many countries, particularly in developing countries, because people in developing countries generally suffer more from injustice and inequality.

Historically, in earlier ages there were certain developments which conferred rights at a national level, such as the Magna Carta (1215), the United States Declaration of Independence (1776), and the French Declaration of the Rights of Man and of the Citizen (1789) (Naseema, 2002). All these historical developments note that human beings have a right to life, liberty, property, happiness, and rules of law. However, similar developments did not occur in the rest of the world. After the Second World War, the international community, especially the United Nations (UN), played an active role in establishing a human rights agenda worldwide by adopting the Universal Declaration of Human Rights (UDHR) in 1948. Basically, the UDHR notes many essential aspects as human rights, such as the right to food, shelter, health, education, property, religion, family life, marriage, and many others. However, it is interesting to note that a 'right to water' was not mentioned in the UDHR and other similar historical documents. Thus, globally, a new issue has emerged. Should water be treated as a human right? The reason is that without water it is impossible to survive.

After the 1980s, water became an important item on the international agenda, because water scarcity is increasing globally because of poor management and water resources are insufficient due to various reasons such as an increasing population, rapid urbanization, and deforestation. For these reasons, many countries have been facing shortages of drinking water and/or water used for industrial purposes or agriculture.

Because of these issues, water has emerged as a topic of major concern in the present world. This paper briefly examines 'water and human rights' as an emerging issue, and also highlights its relation to other areas such as education, health and development.

Objectives and Rationale of the Paper

The objectives of the paper are:

- to examine international efforts, especially by the UN, in terms of water and human rights issues;
- to observe the state–legal realm on water rights law with respect to human rights; and
- to examine the relationship between water, human rights, and good governance, including other water-related issues.

Water is essential and without it it is not possible to survive. Biswas & Tortajada (2006, p. 10) rightly note that 'water is being increasingly considered to be the lifeblood of the planet'. However, the interesting thing is that water is not directly addressed as a human right under the UDHR and other international covenants. Thus, it is challenge in the present world to address the issue of this 'lifeblood' correctly in local/national and international spheres.

Globally, water became a common topic and a significant item on the political agenda since the 1980s. The reason is that water resources are very limited, but demand has increased due to various factors such as urbanization, increased population, industrialization and economic development, and the corresponding higher demands for food, energy and environmental security. These are just a few of the trends that will seriously affect existing water planning, management, and allocation processes (Biswas and Tortajada, 2007). In addition, Biswas (2007a) points out that:

> it is likely that if there will be a water crisis in the future, it will not come because of actual physical scarcity or water, as many predict at present, but because of

continuing neglect of proper wastewater management practices. Continuation of the present trend will make available water sources increasingly more contaminated, and will make provision of clean water more and more expensive, as well as more complex and difficult to manage. By diluting seriously the definition of access to clean water and considering sanitation only in a very restricted sense, developing countries, including many in Asia are mortgaging their future in terms of water security.

Moreover, many countries, international agencies and non-governmental organizations (NGOs) are actively involved in managing water-related problems, especially in developing countries. Singapore is one of the best examples. It has developed extremely efficient demand-and-supply management practices to mitigate its dependence on external sources. Likewise, at the regional level, the Asian Development Bank has been providing support to manage water and sanitation problems in all Asian developing countries such as Nepal, India, Bangladesh, and China. The Bank mostly provides loans for drinking water, sanitation, irrigation, and the energy sectors. However, it gives top priority to drinking water.

The increasingly deteriorating global situation in terms of access to a clean water supply and inefficient sanitation was raised during the UN Conference on Human Settlements, held in Vancouver, Canada, in June 1976. The result of this declaration was that water is considered to be a basic human need (Biswas, 2007b). The Vancouver declaration was then further elaborated by the UN Water Conference, held at Mar del Plata, Argentina, in March 1977. This conference proposed that the period 1981–1990 should be proclaimed as the international water supply and sanitation decade, so that interest in water supply and sanitation could increase significantly at both national and international levels. The recommendation for a decade was subsequently approved unanimously by the UN General Assembly on 10 November 1980. These are some positive signs in the area of water management in an era which gives a high priority to water and directly and indirectly addresses water as a 'Right'.

However, in practice, a fundamental issue is that the water sector is less of a priority agenda in many countries in terms of law and good governance, especially in developing countries. This is because water is not considered to be similar or equal to other rights. For example, many countries have adopted legal provisions such as rights to food, shelter, health, education, employment, life, and liberty. In contrast, many countries lack a commitment to a right to water. Thus, the global scale of the problem creates a huge challenge, and it is both rational and essential in the present context to consider water in relation to human rights.

Limitation of the Study

Water is critical in many areas, that is why it is compared to 'lifeblood'. But drinking water is the most crucial need. Moreover, water is of concern to multiple problems, such as economic growth, an ageing population, urbanization, health (HIV/AIDS), agriculture, industry, energy and environment.

Basic human needs such as food, health, or shelter are now addressed as human rights. However, 'water' is not addressed directly as a right at an international or national level. This paper therefore attempts to examine the present status of water as a human right. It attempts to examine national and international achievements in water and human rights. Moreover, it is based on doctrinal resources and provides an overview of water and related concerns at national and international levels, including international conventions and declarations which consider water as a human right.

Water and Human Rights Framework

General Overview of Water Rights

Human rights and water rights are often used in different frameworks. Basically, human rights cover broad areas, while water rights may vary from one country to another because a country's socio-cultural values impact on individual rights. For example, a one-child policy is adopted in China to control over-population, but Singapore has no such restriction. The next difference is that human rights are natural rights and essential for everyone and everywhere such as the right to life, the right to food and the right to shelter. But a water right is granted by the state to fulfil social needs and maintain law and order. In addition, every state has an obligation to implement human rights without curtailing general norms, but a water right may have some restrictions in terms of utilization such as water rationing in times of scarcity.

If one examines the history of water rights, they emerge from the personal ownership of the land bordering the banks of a watercourse or from a person's actual use of a watercourse or groundwater. Initially, water rights were conferred and regulated by customary values and practices, which are known as common laws in many countries including the United States. Basically, water rights were previously created by contract, as when one person transfers his/her water rights to another by selling the land or waterbody. Such a system still exists in many countries including those in South East Asia, such as Nepal and India. In the Nepalese context, the landowner has first rights to use the watercourse for irrigation, if the watercourse starts from private land (National Code of 1963).

Similarly, in the eighteenth century in America, water was primarily governed by custom and social values. However, rapid population growth and the increasing trend of water use for agricultural and domestic purposes has had an impact on traditional values. For these reasons, water was increasingly considered to be a finite and frequently a 'scarce resource'. Therefore laws were adopted to establish guidelines for the equal distribution of the natural resource. In fact, America has a different legal system from common law, but interestingly American courts also began developing common-law doctrines to accommodate landowners who asserted competing claims over a body of water. These doctrines mainly govern three areas:

- riparian rights;
- surface water; and
- underground water rights.

An owner or possessor of land that abuts a natural stream, river, pond, or lake is called a riparian owner or proprietor. The water law gives riparian owners certain rights to water that are incidental to possession of the adjacent land.

Many international initiatives have been adopted to promote the conservation and efficient management of water use. In particular, the Dublin and Rio Conferences in 1992 were significant in terms of perceiving water as an integral part of the ecosystem, a natural resource and a social and economic good, and in promoting integrated water resources management. The 1st and 2nd World Water Forums in Marrakech (1992) and The Hague (2000) respectively highlighted the role of water, basic human needs, preserving ecosystems, and managing water wisely (Biswas, 2007a, p. 10) and a similar pattern was also highlighted in the more recent World Water Forums.

Water and Human Rights

Water is a basic human need. The fundamental right to life includes the right to drinking water (Rangachari, 2005; Biswas, 2007a). It focuses on the adequate availability and quality of drinking water which is an essential human need because without water, human beings cannot survive. Globally, food, housing, civil, cultural, economic, political, and social rights are regarded as human rights. In practice, human rights ensure fundamental freedoms and dignity, especially for individuals. These rights are in relation to the individual and the state, and the state has an obligation to protect, preserve, and fulfil these rights at the national level.

Moreover, since the 1990s, the human rights concept has been more widely respected and in some areas also come under review by the UN. For example, the Committee on Economic, Social and Cultural Rights (CESCR), which was established to oversee the implementation of the Covenant on Economic, Social and Cultural Rights, presented a document titled the General Comment No. 15 at the UN 29th session in Geneva in 2002. The comment reinterpreted Articles 11 and 12 of the covenant (the International Covenant on Economic, Social and Cultural Rights—ICESCR) and concluded that 'water' can be considered to be a human right under this covenant (Biswas, 2007b).

The importance of a clean water supply and wastewater management became prominent in the international agenda during the UN Water Conference in 1977. The conference proposed that the 1980s be declared the International Water Supply and Sanitation Decade, with the very ambitious objective of providing clean water and sanitation to every human being by the end of the 1990s. Such international efforts have been increasing in the water sector. Thus, the time has come to recognize that people's access to safe drinking water and to sustainable sanitation is a human right.

Furthermore, considering the obligations of the state, it is also important to bear in mind that human beings are responsible for themselves and their own well-being. Human rights do not automatically involve heavy government intervention or imply that individuals can unreservedly demand goods and services from the state. However, current international human rights law is a system of state obligations. Ultimately, each sovereign state is responsible towards those within its jurisdiction, and towards other states and international bodies, for the level of enjoyment of human rights in the country. It is a state's responsibility to regulate the behaviour of all actors within its jurisdiction, to ensure that they respect human rights, and to intervene in case where there are abuses. All human rights entail state obligations, which may be analysed at different levels. ICESCR—1966 Article 2 also stresses a respect for human rights and further emphasizes that states should take legislative, administrative, and other action progressively to achieve the goal that every human being within their jurisdiction has access to adequate water, to the maximum of the available resources. In fact, the primary duty of the state is not to interfere with or deprive people of their rights. This is referred to as an obligation to respect the right in question.

Lastly, the Vienna Declaration and Programme of Action 1993 stressed the rights of everyone to a standard of living adequate for their health and well-being, including food and medical care, housing, and the necessary social services. The World Conference on Human Rights affirmed that food should not be used as a tool for political pressure. Interestingly water is not mentioned in this declaration.

Water, Human Rights and Good Governance

Good governance has been used mostly as an umbrella concept and no definition had been agreed for it as a whole. Governance is not synonymous with government. It is instead a complex process that considers multilevel participation beyond the state, where decision-making includes not only public institutions, but also the private sector, civil society and society in general. Good governance frameworks refer to new processes and methods of governing and changed conditions of ordered rule in which the actions and inaction of all parties concerned are transparent and accountable. It embraces the relationships between governments and societies, including laws, regulations, institutions, and formal and informal interactions which affect all the ways in which governance systems function, stressing the importance of involving more voices, responsibilities, transparency, and accountability of formal and informal organizations associated in any process (Tortajada, 2007).

Likewise, good governance is an important pillar of water management, which enhances various aspects, such as promoting decentralization, building capacity, and strengthening and monitoring evaluation, research and learning at all levels. Further, the governance process helps to mitigate corruption and to make the project transparent and accountable, and it also enhances the provision of more equitable access to water for the poor, especially in developing countries. The poor need to be targeted for equitable access to water. Good governance is critical to reducing poverty, which can facilitate participatory, pro-poor policies, ensure the transparent use of public funds, and promote the effective delivery of public services. The delivery of basic services (such as safe drinking water) matters most to the poor and requires reliable institutional structures and participation by the poor. A long-term objective is, therefore, to empower the poor and develop institutional arrangements that foster participation and accountability at the local level.

Governments should be enablers and regulators of community action for the efficient delivery of water, not water service providers; they should target subsidies to the poorest, strengthen the overall links in the water delivery system and protect the environment. Good governance focuses on policy reform at a national level and encourages decentralized institutions, public–private partnerships, participatory approaches which allow for greater accountability and transparency, and institutional development. All these factors show that water, good governance and human rights are related. In fact, water management is not possible without good governance, and good governance is not possible without respect for human rights. Thus, good governance and human rights both are key pillars for water management.

Water Price and Human Rights

Globally, water became an important item of the international agenda after the 1990s because it is an essential good for all, and a direct link between the livelihood and well-being of humankind. However, water scarcity is increasing. A number of factors impact on the demand for water, such as urbanization, an increasing population, industrialization and economic development, and the corresponding increases in demand for food, energy and environmental security are just a few of the trends that will seriously affect existing water planning, management and allocation processes. To manage all these challenges, water pricing is one solution. However, there are differing opinions on water pricing: some are in favour, but others are against. Those against water pricing argue that if water is treated as a human right, it should be provided without cost. In fact, pricing is a crucial issue in water management, because without water pricing, it is difficult to supply sufficient and clean water to everyone. Therefore, water pricing is important for an equitable water supply.

Limited access to affordable energy and water hinders economic development around the world. By 2025, more than 60% of the world's population will live in countries with significant imbalances between water requirements and supplies, largely in Asia, Africa, and Latin America. More than 1 billion people currently lack access to safe drinking water and 2.4 billion people lack improved sanitation (Atlantic Council, 2005). In this context, governments are not able to manage the problem on their own, so most countries have adopted a public–private partnership programme to manage their water-related problems. Certainly, if the private sector is involved in water management, water will only be available at some cost. Thus, the following two reasons support the pricing of water.

The concept of water pricing and good governance. The *Human Development Report 2006* (UNDP, 2006) highlights the widely held view that each person needs huge amount of water each day for their basic needs—to drink, cook and wash sufficiently to avoid disease transmission (the figure mentioned is 20 litres, but actually its requires about 60 litres). However, in many dry places such as East Africa, people get less than 5 litres a day—in some cases, less than 1 litre a day, enough for just three glasses of drinking water and nothing left for other needs (Black, 2009). Such inadequate water resources make life difficult. Water scarcity results because increasing population, industrialization and economic developments directly demand more water. For this reason, water pricing is a helpful tool for fair distribution and to regulate the water supply.

Japan and Cambodia experience roughly the same average rainfall—about 160 centimetres per year. But whereas the average Japanese person uses nearly 400 litres per day, the average Cambodian must make do with about one-tenth of that amount. These facts show that every country has a different culture and practice when using water and there is no common standard. Therefore, if water is available without cost, there will be considerable misuse. To maintain a logical balance, water pricing is necessary. Thus, water pricing is a powerful tool when the correct costs and incentives are in place to shape public behaviour and the public's consumption of water (Soon *et al.*, 2009, p. 255).

Water price and international instruments. In 2000, the United Nations Committee on Economic, Social and Cultural Rights, the Covenant's supervisory body, adopted a General Comment on the right to health that provides a normative interpretation of the right to health as enshrined in Article 12 of the Covenant. This General Comment interprets the right to health as an inclusive right that extends not only to timely and appropriate healthcare, but also to those factors that determine good health. These included access to safe drinking water and adequate sanitation.

In 2002, the Committee further recognized that water itself was an independent right. Drawing on a range of international treaties and declarations, it stated that the right to water clearly falls within the category of guarantees essential for securing an adequate standard of living, particularly since it is one of the most fundamental conditions for survival. These are core international initiations on water management at an international level, which will help to manage water-related problems in the national realm. However, it is significant that the documents do not mention water pricing although in practice they stress fair and equitable water distribution as a common goal. If water is to be addressed as a common good, it should be priced.

International Legal Framework and Initiations on Water Management

Many countries, NGOs and relevant international agencies have been involved in managing water-related problems by various methods. Singapore is one of the best examples. It has developed very efficient demand-and-supply management practices to mitigate its dependence on external sources. Likewise, at a regional level, the Asian Development Bank has been helping to manage water and sanitation-related problems in many developing countries such as Nepal, India, Bangladesh, and China. The Bank has provided loans for the drinking water, sanitation, irrigation, and energy sectors.

However, the term 'human rights' is used in the sense of genuine rights under international law, where states have a duty to protect and promote those rights in the individual interest. The initial impetus to human rights agreements was to address violations of moral values and standards related to violence and the loss of freedoms. Subsequently the international community expanded rights laws and agreements to encompass a broader set of concerns related to human well-being. Among these are rights associated with environmental and social conditions and access to resources. The extent to which environmental rights are either found in, or supported by, existing human rights treaties, agreements, and declarations has been the subject of a growing literature (Brinie & Boyle, 2002).

For example, Section 27 (1) of the Constitution of South Africa guaranteed that "everyone has the right to have access to ... sufficient food and water". The Housing Subsidy Scheme and the Free Basic Water policy are the South African government's national programmes to deliver on the rights to water and housing. The extent of poor children's access to water and to housing through these interventions are discussed together because access to water is very closely tied to housing or settlement types. In fact, 'basic services', including water, sanitation and electricity, are part of the definition of 'adequate housing' specified by the International Covenant of Economic, Social and Cultural Rights to which South Africa is a signatory and they apply at a national level through the constitutional framework.

The South African government has failed to implement the constitutional provision. For this reason, the Constitution Court decided that the State is responsible for providing adequate housing for all citizens and must create the conditions for access to adequate housing for people at all levels in society. This decision had come about from the Constitution Court in the case of *Government of the Republic of South Africa and Others vs Grootboom and others* (Southern African Legal Information Institute, 2000). Basically, this decision enhanced water management issues in South Africa, including those in the international sphere.

Water and its Challenging Areas

Water and Sanitation

There is a direct relation between one's quality of life and access to water supply and sanitation services. The quality of the water supply and sanitation ultimately contribute towards nation-building and prosperity by enhancing the good health of the people. These are essential requirements for human beings and indirectly relate to human rights.

About 750 million people in rural areas and another 100 million in urban areas still have no access to safe drinking water around the world. Adequate sanitation is needed for 1.75 billion people in rural areas and for 300 million in urban areas (Biswas, 2007b). Until this demand is met, productivity, incomes and health will continue to be impacted, especially for the poor, and the human cost will remain high. Industrial demand continues to rise because water is treated as a social not an economic good and the rapid development and adoption of water-efficient technologies is inhibited. Countries that are beginning to industrialize are witnessing rapid growth in water demand.

In addition, globally, water usage has increased sixfold over the past century, more than twice the rate at which the world's population has grown, and the population growth rate is considered to be high. For this reason, more than 2 million tons of human waste is dumped daily into the world's rivers, lakes and streams, which is a great threat to human health and the environment, especially in developing countries. Moreover, during the next two decades water use is expected to rise by 40%. At least 1.2 billion people currently lack of access to safe drinking water (UNDP, 2006).

In September 2000, at the UN Millennium Summit, world leaders committed themselves to a set of eight time-bound measurable Millennium Development Goals (MDGs). One goal was to ensure safe drinking water and reduce sanitation problems in developing countries. Nearly midway through the decade, about 2.4 billion people (Biswas, 2007b, pp. 35, 42) were living without proper sanitation. The MDG also referred to good governance in water management through government and private partnership.

Despite significant progress being recorded in the extension of basic water supply services across the world, a large section of the population is still suffering from waterborne diseases caused by impure drinking water and insufficient sanitation facilities. The result of inadequate wastewater management is that many countries still have low-quality drinking water; e.g. in Nepal, fewer than 50%; in Cambodia, 53%; in India, 57%; and in Fiji, 87% have access to safe drinking water (Asian Development Bank, 2007). These are some of the challenges for sanitation and safe drinking water which are also critical from a human rights perspective.

Water and Health

In practice, water and health are interrelated because many health problems arise from poor water quality. Safe drinking water is essential for good health, and without safe water, a right to health is not possible.

Human rights and fundamental freedoms are ensured in some international covenants and declarations. The right to health is also considered in the covenants. For example, Article 25 of UDHR stresses that everyone has a right to a standard of living adequate for the health and well-being of himself and of his family, including food, clothing, housing, 'medical care', and the necessary social services. Likewise, Article 12 of ICESCR also recognizes the right of everyone to the highest attainable standard of physical and mental health.

In addition, the World Health Organization considers that the highest attainable standard of health is one of the fundamental rights of every human being (WHO, n.d.). Further, the Convention on the Rights of the Child (CRC) 1989 states that children are entitled to the highest attainable standard of health, which requires state parties to take appropriate measures to combat disease and malnutrition, including within the framework of primary healthcare. Beside this, the World Health Organization stresses that health is one of the fundamental rights of every human being and should be emphasized through various programmes at a national level.

The right to health, a decent existence, work and occupational safety and health, the right to an adequate standard of living, freedom from hunger, an adequate and wholesome diet, decent housing, the right to education, cultural equality and non-discrimination, dignity and harmonious development of the personality, the right to security of person and of the family, the right to peace and the right to development are all rights established by existing UN conventions. These rights represent the ideal that governments strive for in providing for their citizens' basic life requirements that all humans are entitled to (Johnston, 1996, p. 7).

Water and Education

Education and awareness are crucial for water management which aims to promote wide-ranging public awareness and community education programmes especially among women, youth and farming groups to convey the message that water is a resource that needs prudent management.

Water education is useful to create awareness in public life, e.g. among people in rural areas, especially women, illiterate farmers and the younger generation, as well as children. Basically, education is beneficial to mitigate water-related problems such as water pollution. Likewise, if academic institutes introduce water education courses in school, college or university, it will promote water awareness at an academic level. Actually, such developments will have a large impact in terms of water management. For example, the Lee Kuan Yew School of Public Policy in Singapore has initiated a 'Water Policy and Governance' course in its graduate programme. The courses will have a positive future impact in the region. In addition, the trend for educating a younger generation in the water sectors has been increasing in many countries, especially at university level.

In addition, the Asian Development Bank also promotes public awareness and community education programmes especially among women, youth, and farming groups. Education particularly helps communities understand the linkages between water, sanitation, health and productivity. Clearly, water helps to change mindsets. It is insufficient for policy makers only to approach the management of water resources differently. Those who consume water also need to learn through education to recognize the critical nature of the resource.

Water and Stakeholders

The challenges of ensuring that water is conserved and managed wisely are huge and no single agency can hope to address this area alone. Strengthening partnerships among stakeholders will be crucial for policy implementation. Principally, these are necessary for ensuring cooperation between various agencies such as governments, the private sector, NGOs, and donors. Such cooperation will be factored into the action agendas and stakeholder partnership agreements established to foster a sense of accountability. Stakeholder partnerships with the development community should be sought to maximize the impact of external resources. The partnership approach will clearly identify responsibilities in terms of legislative change, policy reform, institutional change, capacity building, and the financing of high-priority investments. This coordination will be at country, regional, and global levels and will cover operational, sector, and awareness-creation work.

Moreover, the promotion of participation involving public, private, community, and NGO stakeholders is a key element of this policy. The outcome will depend on the existence of laws, regulations and policies to regulate water-sector activities and their fair, equitable and consistent application.

Water and Gender Issues

Women and children are more vulnerable to the effects of water scarcity in the South Asian developing countries such as Nepal, India, and Bangladesh, especially in rural areas. The main reason is that traditionally and culturally women are more involved in housework such as cooking and cleaning, which means that they engage more in water-related activities. Likewise, women are more engaged in fetching drinking water from faraway places, which is an extra burden for women in the family. It means that men are less responsible and worry less about fetching water than women. This stereotypical socio-culture has created many barriers between men and women, for example, in terms of women's employment and education. However, since 1970, the worldwide feminist movement has increased the breakdown of such a stereotypical/discrimination role to create a greater gender balance in the public and private spheres. The UN has especially played a significant role in promoting gender equality in various areas including public and private matters in the water sector.

If a gender balance is maintained in society, it will also reduce the present gender imbalances in the water sector. However, while gender issues and solutions in water supply, sanitation, and hygiene are comparatively well researched and implemented, good practices in connection with water and land rights and in the management and conservation of resources have not been widely adopted (Biswas, 2007b; Asian Development Bank, 1998).

Water and gender issues are gaining prominence, and in this regard the Asian Development Bank's water sector operations policies have incorporated some key elements in a gender approach to planning, implementing and evaluating water sector activities, which are:

- including a gender analysis at the design stage;
- incorporating explicit gender-equality provisions in the objectives and scope of the activity; and
- disaggregating data in monitoring and management information systems along gender lines.

Water and Food/Hunger

Water scarcity reduces food production in many countries. If less food is available, there will be increasing levels of hunger in coming years. In this respect, water, food and hunger are connected to each other. Hunger creates beggars, robbers, environmental pollution, displacement, poverty and crime. These situations in turn create more social factions and instability within society. Currently, more than 850 million people around the world are facing hunger every day, half of them in Asian countries. Far more people die from causes related to chronic hunger than to famines. Less developed countries within Asia suffer more from hunger, and the number is growing steadily. It is a serious problem for nations. Globally it is estimated that nearly 226 million children are stunted due to hunger-related problems (Friends of the World Food Program, n.d.). Thus, the coming years will present a challenge for combating hunger due to water scarcity, particularly in irrigation and agriculture production.

Water and Management Skills

In those South Asian countries which have developed least, such as Nepal and Bangladesh, the economy is based on agriculture and water-related developments, especially crops and livestock. In practice, these countries have huge water resources. For example, Nepal is regarded as a 'water-rich' country because it has more than 6,000 rivers, and the total length of the river courses is estimated at 45,000 km. The total (90% reliable) annual flow of major rivers of Nepal is estimated as $1.25 \times 10^9 \text{m}^3$ (Chatterji *et al.*, 2002; Alford, 1992). There are also groundwater resources such as the deep and shallow aquifers in the Terai and Siwalik hill valleys. However, spatial and temporal distribution creates surpluses as well as shortages within the country. It means that not water availability but the management of water resources is the core problem for most developing countries. Weak management skills have a negative impact on water development in many developing countries, including Nepal.

In the future, water resources and the demands placed upon them at the local and regional scales will become increasingly complex and disparities in access to water resources are likely to grow. On the other hand, many developing countries have adopted some key instruments for water-sector development, e.g. decentralization processes, private sector participation, and public–private partnerships. These processes are likely to improve water management. For example, in the Philippines, water supply and sewerage services in the Metro Manila area were awarded in 1997 as concession contracts for 25 years (Biswas, 2007b). Service quality has improved markedly with regular hours of supply, fewer interruptions, an improvement in water quality, and significantly reduced water bills. These show that new management skills and methods are also needed for water-sector development.

Conclusion

Water is essential for human beings, whether it is free or priced. However, the bigger question concerns the quality and sufficient supply of water. In recent years water has become an important issue in nearly all developing countries.

Globally, the International Covenant on Economic, Social and Cultural Rights (ICESCR) has been able to address water as a 'human right' under the international framework. Likewise, some developing countries such as South Africa are also able to address water as a human right under their constitutions. In addition, since the 1990s, the global trend has rapidly changed because most countries have adopted privatization, globalization, and public–private partnership programmes. These reforms also impact on the water sector, where many developing countries have adopted private-sector involvement to improve water delivery services. Whether water services are provided by the public or private sector, water has to be priced to ensure it is properly used by the users, and also that utilities have incomes with which they can at least cover their operation and maintenance costs.

Declaring water as a human right will not solve the world's water and wastewater management problems. However, it is a step in the right direction. What is needed is a new mindset among the water professionals and policy makers which will allow the world's water problems to be solved. There is no question that the world has enough water. What is needed is more efficient water management.

References

Alford, D. (1992) *Hydrological aspects of the Himalayan region, ICIMOD Occasional Paper No. 18* (Kathmandu: ICIMOD).

Asian Development Bank (1998) *The Bank's Policy on Gender and Development* (Manila: Asian Development Bank).

Asian Development Bank (2007) *Asian Water Development Outlook* (Manila: ADB).

Atlantic Council (2005) *A Marshall Plan for Energy and Water Supply in Developing Countries*. Available at http://www.acus.org/program-energy_projects-Marshall.asp/.

Birnie, P. & Boyle, A. (2002) *International Law and the Environment* (Oxford: Oxford University Press).

Biswas, A. K. (2007a) Water as a human right in the MENA Region: challenges and opportunities, *International Journal of Water Resources Development*, 23(2), pp. 209–225.

Biswas, A. K. (2007b) Water for all. Paper presented at the Asia-Pacific Water Forum, Asian Water Development Outlook, 2007, p. viii (Manila: Asian Development Bank).

Biswas, A. K. & Tortajada, C. (2006) Changing global water management landscape, in A.K. Biswas, C. Tortajada & R. Izquierdo (Eds) *Water Management in 2020 and Beyond* (Berlin: Springer).

Black, R. (2009) Water – another global 'crisis', *BBC News website*, available at: http://news.bbc.co.uk/1/hi/sci/tech/7865603.stm (accessed 21 March 2009).

Chatterji, M., Arlosoroff, S. & Guha, G. (Eds) (2002) *Conflict Management of Water Resources* (Burlington, VT: Ashgate).

Friends of the World Food Program (n.d.) Available at www.friendsofwfp.org.

Johnston, B. R. (Ed.) (1996) *Society for Applied Anthropology 'Who Pays The Price? The Socio cultural Context of Environmental Crisis*, p. 7 (Washington, DC: Island).

Naseema, C. (2002) *Human Rights Education, Conceptual and Pedagogical Aspects* (New Delhi: Nice Printing Press).

Rangachari, R. (2005) *Bhakra-Nangal Project, India: Regional and National Impact* (New Delhi: Indian Water Resources Society).

Roemer, R. (2000) El Derecho a la Atencion de la Salud, in: H. L. Fuenzalida-Puelma & S. S. Connor (Eds) *El Derecho a la Salud en las American*, Publ. No. 509, p. 16 (Washington, DC: OPAS).

Soon, T. Y. with Jean, L. T. & Tan, K. (2009) *'Clean, Green and Blue', Singapore's Journey Towards Environmental and Water Sustainability*, p. 255 (Singapore: ISEAS Publ. Institute of Southeast Studies).

Southern African Legal Information Institute (2000) *Government of the Republic of South Africa and Others v Grootboom and Others*, Available at: http://www.saflii.org.za/za/cases/ZACC/2000/19.html.

Tortajada, C. (2007) Paper presented at the 2nd Partners Forum 'Water Governance in the MENA Region: Critical Issues and the Way Forward', organized by Went Capacity Building International in Germany and the Arab Water Council in Cairo, 23–26 June 2007.

United Nations (1948) *Preamble of the Universal Declaration of Human Rights* (New York: United Nations).

United Nations Development Programme (UNDP) (2003) *Clean Water and Sanitation for The Poor*. Available at: www.undp.org/.

UNDP (2006) *Human Development Report: Beyond Scarcity: Power, Poverty and the Global Water Crisis* (UNDP).

WHO (n.d.) *WHO Constitution*, Available at: http:www.who.int/hhr/en/.

Terrorism—A New Perspective in the Water Management Landscape

TRISTAN SIM TONG PING

Lee Kuan Yew School of Public Policy, National University of Singapore, Singapore

ABSTRACT   *This paper establishes the urgency of water terrorism by showing how the drinking water system can be and has been attacked. The paper discusses the lack of global dialogues on water terrorism and urges officials and experts to rethink the issue. Even as global water experts engage in deep discussions to improve water systems and accessibility, such dialogues, while seemingly unrelated to terrorism, can indeed be starting points for discussions on water terrorism. A simple approach to protect the water system is suggested. It is impossible to protect against every known threat, but there are many cost-effective solutions. By applying some basic, pragmatic guidelines, even developing countries can adopt cost-effective solutions to protect their water systems.*

Introduction

Global water experts are concerned with troubling trends such as urbanization and population growth; institutional concerns such as mismanagement and corruption; and systemic problems such as pollution, poor infrastructure and supply inadequacies. They are trying to lay more pipes and build better facilities to treat and supply better quality water to more people.

Policy-makers have their plates full as they grapple with the above issues—issues that are subject to widespread and on-going debates. Progress is being made as more countries take a good, hard look at the worsening trends. However, there is one particular uncertainty that can cause unprecedented disruption and deaths, whose effects can completely overshadow all the current issues, yet which is receiving scant attention. That uncertainty is water terrorism.

There is no universally accepted definition of terrorism. By defining the term in their constitutions, governments and international organizations such as the United Nations define the scope of their roles, accountability and responses against terrorist acts. Nonetheless, as a working definition this paper uses the term 'terrorism' as defined by the US Department of Defense:

> The calculated use of unlawful violence to inculcate fear, intended to coerce or to intimidate governments or societies in the pursuit of goals that are generally political, religious, or ideological.

No country in the world is immune from terrorism. Where political, religious or ideological differences exist, there is a potential for human conflicts which may result in terrorist acts. Water system security is a low priority in many countries, yet it is precisely because of the great daily dependence people have on water that it is such a vulnerable and attractive target. Water terrorism is a real threat and a terrorist has a host of ways to attack the water system, from using biological and chemical agents to physical attacks on water companies, treatment plants, reservoirs and dams to disrupt and contaminate the water (Gleick, 2000; Meinhardt, 2005; Burrows & Renner, 1999; Zirschky, 1988; Khan et al., 2001).

The good news is that the latest treatment processes such as reverse osmosis filtration, ozone, ultraviolet treatment, and chemical treatments can take out the bulk of the biological and chemical agents. Also, with large-source water flows as in major rivers, the dilution effect can nullify the chemical and biological threats. The bad news is that in some countries the treatment plants need repairs, upgrades or replacement, and the pipes, distribution and storage networks are old and crumbling and easily accessible.

For example, in Dushanbe, Tajikistan, when the municipal water supply stopped using chlorine due to a lack of funds, the untreated water caused approximately 8,900 cases of typhoid fever and 95 people died from it over a two-year period from 1996 to 1997 (Khan et al., 2001). If terrorists were to exploit such weaknesses at source, they could cause great harm on a large scale.

Many countries have ancient water infrastructures that are simply not built with terrorism in mind. But all is not lost. Even as governments respond to water experts' dire predictions of the looming water shortage crisis, they can start to rebuild the water infrastructure system and incorporate security designs to withstand or deter terrorism. For existing structures that are in good condition, simple, cheap and common-sense security improvement can be made to shore up the defences.

The Threat of Water Terrorism

Absence of a Global Focus

Threats from water can come in the form of man-made and natural disasters such as chemical spills, floods, and tsunamis to more mild occurrences such as sewer leakage or runoff from agricultural lands. Water can also be used as a deliberate means of delivering terrorism. However, water-related terrorism is not accorded much weight or urgency in the current literature and global debates on water issues.

Water experts have relentlessly debated over water management policies. The debate is taken up by world bodies such as the United Nations, the World Bank, the Asian Development Bank, and many other notable non-governmental organizations (NGOs).

What the global water bodies are concerned with is the management of water supply and demand. For example, the United Nations Children's Fund (UNICEF) and the World Health Organization (WHO) jointly reported that 884 million people worldwide are without adequate drinking water, and 2.5 billion are without adequate water for sanitation such as wastewater disposal (UNICEF/WHO, 2008). The World Bank estimates the number of people who lack access to clean water to be closer to 1.1 billion (Oliver, 2007). Whatever water they have is badly contaminated, highly priced or difficult to access, making their living conditions miserable, even as many are struggling with poverty.

Water experts generally agree that the scale of water-related problems is huge and the trends, if not arrested, could grow into a water crisis within the next 20–30 years. In the *Asian Water Development Outlook 2007* report (Asian Development Bank, 2007), Biswas pointed out that:

> It is likely that if there will be a water crisis in the future, it will not come because of actual physical shortage of water, as many predict at present, but because of continuing neglect of proper wastewater management practices. (p. 16)

The neglect of wastewater management is a decades-long problem which could build into a water crisis in the future. Now, imagine the contamination rate of water multiplied many hundred-fold in a single instance of time so that the water crisis is immediate. That is the kind of threat represented by water terrorism.

The two tracks of water management and water terrorism have not collided in the international arena. Water experts have not thought much about linking the global water debates to terrorism. When 'water security' is mentioned, it usually means securing the sources of water rather than securing against a terrorist attack. The only mention of terrorism in the Millennium Project was that 'Far more people endure the largely preventable effects of poor sanitation and water supply than are affected by war, *terrorism*, and weapons of mass destruction.'

For water terrorism to be taken seriously as a candidate for global discussion, it must be shown to be relevant to the ongoing debate in the global water management landscape. The discussion below attempts to identify how global water experts could consider the threat of terrorism meaningfully.

Potential Failure of the Water Delivery System

To understand how to prevent terrorist attacks, it is necessary to understand how such attacks can occur. Courtney (2003) states three major objectives for a water delivery system: consumption, public health, and fire-fighting. Consumption is about making clean drinking water accessible, free from chemicals, diseases, and other toxic contaminations. Public health is concerned with the proper treatment of human and industrial sewerage. Lastly, fire-fighting requires ensuring access to fire-hydrants that have sufficient pressure to put out fires.

Those living in relatively well-developed countries such as Singapore take clean water for granted. When the tap is turned on, clean and drinkable water is available any time of the day. When the Singapore Civil Defence fire-fighters rush to an incident, they can easily hook up to a fire-hydrant and expect water to pump through at a rate more than sufficient to handle any fires.

In a nightmare scenario, all three objectives could fail at the same time. A terrorist attack could contaminate the drinking water in storage tanks, disrupt the water supply with well-timed explosions, and simultaneously start fires at chemical factories and residential buildings. In a densely packed city, this would quickly become a major crisis resulting in widespread panic when the tap water is no longer safe and the price of bottled water has skyrocketed. Fire-fighters would have limited capabilities due to the disrupted water supply and low water pressure for fighting fires, so they would be unable to protect homes and save lives. Chemicals and petroleum-related industries would give off toxic fumes as they continued to burn. At night, the city would be thrown into darkness as power generation plants shut down due to the disruption of the water supply. This, in turn, would affect the operations of water utilities which rely on electricity.

Whilst an unlikely scenario—mounting such an attack would require considerable research and resources—the inherent weaknesses in many water systems could be exploited by terrorists as specific targets. For example, in many urban cities large numbers of residents live in high-rise buildings and water is distributed by pumping it to the storage tanks on the rooftops of these buildings. The gates to the water tanks are usually secured with simple locks and there are no electronic monitoring devices to alert the authorities to any tampering. Furthermore, the tanks are easily accessible by contractors who maintain them and it would be a relatively simple task to introduce biological or chemical contaminants into such tanks. All that would be needed is a heavy-duty cutter, some contaminants, and a new lock to replace the old one. Once the public realized that some tanks were contaminated, nobody would trust water from the taps anymore.

Types of Threats

In the Bible, with knowledge of the looming threat of the great flood, Noah built an ark to save his family and some chosen animals. Similarly, the first step to protect and safeguard the water system is to have a basic appreciation of the threats of terrorism so that water practitioners can take appropriate measures to protect this valuable resource. The range of common threats faced by water systems all over the world includes the following:

- Contamination threats through chemical and biological contamination.
- Cyber threats, which include the control and disruption of the computerized network Supervisory Control and Data Acquisition (SCADA) system.
- Physical threats such as a bomb attack or sabotage on a facility (Marta, 2006).

Foran & Brosnan (2000, p. 993) believe that 'terrorist use of bio-weapons poses a significant threat to drinking water. Several pathogens and biological toxins have been weaponized, are potentially resistant to disinfection by chlorination, and are stable for relatively long periods in water'. By contaminating water supplies at source or in the distribution and storage system, terrorists can make use of an efficient water connection system to strike fear and panic into homes and offices directly and quickly.

A terrorist could choose to attack the water system at many points. Meinhardt (2005) suggests the following weaknesses:

- Upstream water sources such as rivers, streams, reservoirs and dams.
- Water treatment plants.
- Distribution systems to neighbourhoods, hospitals, schools, etc.
- Home and office water connections and storage tanks.
- Bottled water and water used for food processing.
- Recreational water such as swimming pools and lakes.

Knowing the potential weaknesses of the water system is the key to changing the mindsets of officials to design effective defences against attacks. The next section establishes the threats.

International Terrorist Trends

For global water experts to see the urgency of installing security measures to protect water systems, evidence is necessary to show how 'real' water terrorism is. Contrary to widespread belief, data on terrorist activities show that there have been numerous water-related terrorist acts throughout history. Gleick (2000) lists 52 acts of water terrorism between 1748 and 2006. The following examples illustrate some of the more recent cases (Gleick, 2000):

- In 2000, a hacker, fired from his job, caused a waste treatment plant to discharge raw sewerage into parks and creeks in Queensland, Australia.
- In 2001, Italian police arrested four Moroccans with ties to al-Qaeda allegedly planning to contaminate the water supply system in Rome with a cyanide-based chemical, targeting buildings that included the US Embassy.
- In 2001, Philippine authorities were forced to shut off water to six villages after complaints of a foul smell from the tap were received. Abu Sayyaf terrorists had earlier threatened to poison the water supply.
- In 2002, the FBI sent out alerts to water managers that al-Qaeda were scouring the Internet for information related to the SCADA computer system that controls and runs the water supply and treatment plants.
- In 2003, al-Qaeda threatened the US water system stating that it does not 'rule out ... the poisoning of water in American and Western Cities'.
- In 2006, Tamil Tigers cut off water to some villages in North Eastern Sri Lanka.

Not only are there still sporadic cases of water terrorism all over the world, but there are also many contaminants that terrorists can acquire. Khan *et al.* (2001) listed 16 biological agents which can constitute a water threat. The different biological agents can stay stable in the water supply for some time, ranging from 2 days to 2 years. A biological agent can be a microscopic parasite, toxin, virus, fungus, or prion. Some examples include: *Bacillus anthracis* (a bacteria that causes the lethal anthrax disease); *Salmonella* (a bacteria that causes typhoid); hepatitis A (a virus that attacks the liver); botulinum toxins (a highly lethal protein toxin, also known as botox); and ricin (a protein toxin from castor beans). Khan *et al.* (2001) also provide further examples of water terrorism:

- Kurdish rebels who tried to use potassium cyanide to poison a Turkish military base in 1992.
- A German biologist who tried to blackmail the government for US$8.5 by threatening to contaminate water supplies with *Bacillus anthracis* and the botulinum toxin.

Economic Cost

Not only can terrorism cause morbidity and mortality, but it can also inflict severe economic costs. To terrorists, the attacks yield spectacularly high returns on their investments, which enhances the attraction and provides an incentive to continue.

Lederer (2004) cited a United Nations report which stated:

> The al-Qaeda terror network spent less than US$50,000 (A$71,000) on each of its major attacks except the September 11 suicide hijackings and one of its hallmarks is using readily available items like mobile phones and knives as weapons.

In the United Kingdom, the bill for terrorist attacks was in the hundreds of millions pounds sterling over the past two decades. The September 11 attacks in the United States cost the nation more than US$100 billion, which includes life insurance, damage to property, and the loss of production.

While water terrorism is rare compared with conventional terrorist attacks such as the bombing of iconic or soft targets (e.g. the transport system, embassies, or busy tourists spots), the cost of a water-related terror attack can be very high. In Milwaukee, Wisconsin, about 403,000 people were affected by one of the biggest water-borne disease outbreaks in 1993, resulting in widespread diarrhoea. A total of 52% of the population receiving water from the municipal water treatment plant were affected, even though the plant had met all existing standards. A total of 4,400 people had to be hospitalized, and at least 54 died. The culprit was cryptosporidiosis (*Cryptosporidium* is a parasite that attacks the intestines), which managed to get through the water treatment system. The cost to the economy was 725,000 productive-days lost, costing more than US$54 million in lost work and expenses (Meinhardt, 2005). While this was not a terrorist attack, it shows how easily a contaminant can harm a local population and the economy.

Alarming but Not Alarmist

While this paper aims to establish the threat and urgency of water terrorism to spur international discussions on the issue, the debate on water terrorism should not be raised to alarmist proportions since this could backfire.

Petrasek (2003, p. 6) stated that 'terrorist threats often are more perceived than real risks'. The mere suggestion that the water is contaminated can cause widespread fear as the public can become overly concerned very quickly, even if the risk of contamination is low. Add media sensationalism to the mix and a city will end up with mass panic and pandemonium. To have any chance of asserting a sense of control and calm during a terrorist threat situation, water practitioners will need to have good water-security measures in place so they can authoritatively reassure the public (Hamm, 2001).

Gleick (2000, p. 500) suggested that:

> Among the best defences against terrorist threats to water systems are public confidence in water management systems, rapid and effective water quality monitoring, and strong and effective information dissemination. New tools for communicating with water users may be valuable in countering the threat of water-related terrorism and ensuring public confidence and calm. Such tools will also have value during natural disasters and accidents.

When water practitioners have better information about the real threat of water terrorism and understand the urgency of implementing security measures, it is the hope of this paper that they will follow through with real actions. Behaviours can only change when the hearts and minds of practitioners have been won. This can be done not by sounding alarmist, but by properly forming a logical case.

This paper advocates a pragmatic, conscious and deliberate approach to the threat of terrorism. Firstly, it is necessary to acknowledge that the threat exists—there are jihadist, extremist, fundamentalist, self-radicalized locals and all sorts of labelled madmen who can start a terrorist attack in any country, developed or developing. Secondly, governments can prepare plans to deal with the aftermath of terrorist attacks, similar to plans for natural disasters such as hurricanes, earthquakes, or floods that can also damage water utilities. Thirdly, governments can find the resources physically to strengthen facilities to deter and protect against attacks. The next section provides a framework to protect the water system.

Protecting the Water Systems

Librarians were shocked and appalled when the federal government told them to destroy all copies of a CD containing a U.S. Geological Survey document that contained information on public water supplies. (Doherty, 2002).

Drastic times call for drastic measures. What do countries need to do to protect their water systems? In many developing countries, the utilities have enough problems coping with old systems and a failing infrastructure. But the threat of terrorism simply cannot be ignored as lives, long-term public health, and environmental damage can result from acts of terrorism.

The Need for New Paradigms

Biswas & Tortajada (2009) argue that planning for future water developments using current principles and paradigms will fail. The water needs of the future are becoming increasingly complex and interlinked—arising from competing and increasing demands, each with its own set of issues to consider. Even if water officials want to do their best, applying their past know-how to future projects will doom them to failure.

Applying this philosophy to water terrorism, if current water practitioners have a deeper understanding of water terrorism, they too will be able to *learn from past mistakes* such as not securing water facilities, *adapt to current needs* of security in a post-9/11 world where terrorists are still planning the next attack, and be better able to *anticipate the future threats* so that better plans can be made.

As modern infrastructures and facilities are planned and built, officials should consider incorporating security protocols. Without adequate planning, future water systems will not be able cope with growing demands. Without adequate provisions made for terrorist-related measures, future water systems will be like 'sitting ducks' waiting to be targeted by terrorists.

Technology

The improvement in technology is seen as a positive trend in the global discussion on water management as it helps bring down the cost of treating raw water. But when the element of terrorism is added, the issue takes on a darker turn. The concern and fear is that the threat may get worse as technology has created more opportunities for terrorists. For example, there is a risk of terrorist agents hacking into the SCADA control system or using special microcapsules to disperse human pathogens into drinking water systems (Foran & Brosnan, 2000).

Water plants employ techniques to ensure clean drinking water. General filtration (excluding reverse osmosis and ozone) coupled with chlorine treatment can kill most bacteria, but they are ineffective against toxins, chemicals, and some parasites

(Khan *et al.*, 2001). While activated carbon can remove some organic toxins, not all water plants use the system. As water filtration and treatment technology improves and costs come down, more can be done to ensure the safety of drinking water.

Technology-based and other pre-event or pre-exposure management strategies can be effective deterrents to contamination threats. New and developing technologies will enable utilities to detect pathogens rapidly in real-time, both in source water and in water distribution systems. Included among these emerging technologies are DNA microchip arrays, immunological techniques, micro-robots, and a variety of optical technologies, flow cytometry, molecular probes, and other techniques (Foran & Brosnan, 2000).

Researchers at the University of Wisconsin–Milwaukee's Centre for Water Security are developing various new tools to detect and provide early-warning of contamination in the water supply so that water utilities can react swiftly to possible threats (Mader, 2003). Some of the research includes:

- Organisms as detection systems, such as special zebra-fish with firefly DNA that glows when toxic chemicals are present, and laser-illuminated camera systems to detect changes in behaviours of microscopic zooplankton and Zebra mussels toward harmful materials.
- Fibreoptic-based sensor networks that will monitor whether a potent toxin such as ricin is present, how quickly it is moving and where it is located.
- Real-time atmospheric and lake modelling systems that determine where, how much and how rapidly a contaminant spreads. Calculations will then be used for emergency planning and strategies.

The Price of Security

The issue of how to price water correctly has frustrated many water experts over the years, especially when they see wanton wastage and abuse through improperly priced systems. The inclusion of the terrorist dimension is set to complicate matters further. Protective measures cost money. Depending on the security requirements, the cost of protecting a water facility can vary greatly—from a cheap CCTV surveillance system to a more expensive system that allows live monitoring of hazardous chemicals in the water.

While it is usually dependent on the state to fund enhanced security measures in the form of vulnerability studies and physical upgrades, there are other downstream costs, which can include recurrent operations and maintenance. For example, extra security officers will have to be paid and CCTVs will have to be maintained regularly for peak performance and eventually replaced.

Water utilities will also have to strike a balance between security spending versus capital spending for maintaining the infrastructure, adding capacity and complying with regulations (Wade, 2002). Utility companies over time will need to factor in security fees to water bills to recover security costs. This is similar to airports charging for security fees when people travelled by air after 9/11.

So far, this paper has established a credible water terrorism threat and suggested the need for a fresh paradigm concerning the planning for future water security, as well as due considerations of the available technology and sustainable funding costs. Water managers should next consider the security measures to implemented.

Six-Step Security Measures

Terrorists are opportunistic target-seekers. They want to attack a target and be confident of success. They do not want to fail. Even adding a few patrol guards around a facility can make a terrorist think twice and reconsider his plans accordingly. Deterrence is the name of the game when most governments embark on counter-terrorism measures.

It is commonly acknowledged among terrorist experts that it is impossible to stop a determined suicide-bomber whose intention is to kill innocents and destroy public property. He can easily access a crowded subway or shopping mall and activate a bomb. But many governments continue to install CCTVs and deploy security officers to patrol such areas, with the aim of deterrence. Before an attack, a terrorist needs to survey the area and rehearse in order to time his attack for maximum impact. Security measures such as guards and CCTVs act as layers of detection and a deterrence to such would-be attackers. The more security layers that are implemented, the higher the chances of detection and the greater the deterrence to the attacker.

The concept can similarly be applied to water facilities. Utilities in developing and developed countries start from different bases of security needs and, hence, require different levels of security measures and funding. The amount of government funding will also determine what a water manager can do to enhance the security of his plant.

This section aims to provide some basic technical information on how water infrastructure can be protected, how resources can be prioritized and the basic focus of protection. It is not meant as a comprehensive security guide, but rather to highlight for water professionals a simple and basic approach to securing water systems. The help of security professionals should be sought for any serious security upgrades or plans.

A simple six-step process is proposed:

- vulnerability assessment;
- risk management;
- strengthening existing infrastructures;
- security by design;
- contingency plans; and
- exercises.

Vulnerability Assessment

The water and wastewater system components such as reservoirs, pumps, pipelines, storage tanks and maintenance centres are vulnerable to terrorist attacks because of the extensive network of unprotected and accessible parts. Physically destroying a part of the system rather than contamination is usually more readily accomplished, but it is important to consider all threats and vulnerabilities within the system.

To assess the vulnerability of the system, a security evaluation should be made of the potential threats to people, equipment, and facilities. Often done by professionals, the vulnerability assessment focuses on the activity rather than on the attacker as 'attackers can come in many forms—from employees and vandals to renegades and terrorists—but threats to the water system are limited to what those people can do' (Wade, 2002).

The US Environmental Protection Agency (USEPA) defines the process of a 'Vulnerability Assessment' for water utilities, as a systematic analysis used to develop a security protection plan for water supply, treatment, and distribution systems. The Assessment identifies a system's vulnerabilities and provides a prioritized plan for security upgrades, modification of operational procedures, and policy changes to mitigate identified risks to critical assets. It also provides a basis for comparing the costs of protection against the risks posed (Courtney, 2003).

The vulnerability assessment is based on the three major categories of threats: physical damage, contamination, and cyber attacks:

- Physical damage: a terrorist can damage facilities or equipment such as pumps, pipelines, and chemical storage tanks. Chemical storage includes tanks containing chlorine gas, caustic soda (sodium hydroxide), and orthophosphate (phosphoric acid). The assessment examines how the various areas are protected and whether redundancies are built in.
- Cyber attack: many water systems are automated and managed by supervisory control and data acquisition (SCADA) systems. These may have vulnerabilities that will allow a hacker to control the SCADA system either locally (by gaining access to the centre) or through the Internet (by penetrating security firewalls). While most systems can be operated manually in case of an attack, 'The real question is: Does anybody know how to do it? When you build a new plant and everything runs by computer, do the operators ever simulate manual operations?' (Wade, 2002).
- Contamination: the public are concerned about biological, chemical, and even radiological agents contaminating the water supplies. However, most critics are of the view that the threat is very low as it would take trucks full of contaminants to threaten the safety of the water system seriously (Wade, 2002). But this cannot be taken too lightly. Firstly, due to the very public nature of fear it can arouse; and secondly, the public health symptoms may only emerge after a period of days (Courtney, 2003).

Risk Management

All managers want the best safeguards available, but have limited funds with competing needs for daily operations, future upgrades, and now security. The vulnerabilities highlighted above do not equate to the risk of actual occurrence. 'Risk' is determined by the probability of an attack happening, while 'threat' is what the enemy can potentially use to attack the water system. Risk must also be evaluated based on the potential consequences to people and property.

Risk management is about getting the most out of limited funds by prioritizing security measures. Since it is impossible to protect everything, risk management forces the manager to think about what level of risk is acceptable and which are the most critical areas to be protected first.

For example, given US$100,000, a utility manager may prioritize the spending on first enhancing the access controls at the entrances to the facility, improving the surrounding lighting, then installation of CCTVs at the perimeter, and finally motion sensors at the fence-line if the budget permits.

Strengthening Existing Infrastructures

Most of the strengthening measures involve putting in layers of protective rings in a given infrastructure. The inner ring is concerned with the access controls (entrances and exits) leading into the utility. This can range from photographic personal security passes to state-of-the-art biometric scanners. The middle ring can be based on perimeter security, which includes anti-climb fences with razors on top, motion-detection devices, and mechanisms to alert operators. The outer ring can comprise active guard patrols on the ground and CCTV scanning of the exteriors. Finally, early-warning monitoring systems can form the outermost ring. Such systems can reliably identify contamination events in the source water or distribution systems, so that response measures can be activated. This is probably the most costly part of the defence.

Apart from installing strong doors and locks, it is important that employee security awareness is raised. Many times, complacency sets in after a period of inactivity and that is when the terrorist can strike. The community must also be roped in to act as eyes and ears so that any suspicious activity or event can be quickly reported and investigated.

The most effective deterrents are often common-sense measures. Some of the USEPA advice includes keeping gates locked, ensuring fences are maintained and that there are no trees or overgrown grass around the perimeter, and increasing lighting at night. Utilities should issue security badges to staff and require everyone entering facilities to have either a badge or a staff escort. It should be difficult for anyone inappropriate to enter. Staff should be trained to be more aware of anomalies in or near sites (Petrasek, 2003).

Designing Security into Future Infrastructures

Even as water planners design modern facilities that can cater to increasingly large populations, they should design in security from the start. Security often occurs as an afterthought in architectural designs of facilities. To then retrofit the facility with security measures will be extremely costly, inefficient, and disruptive to employees. When the architects, facility manager, and security expert get together at the drawing board stage to design a new facility with security in mind, the outcome is much better. Blind-spots and vulnerabilities can be designed out, making it that much harder for a terrorist to strike successfully.

An example of how better designs can be incorporated is as follows: there will be more setback (to increase the distance from a potential car bomb on the roadside to the utility), fewer entrances (to make it easier to monitor and protect), fewer numbers of glass facades (to minimize secondary fragmentation of glass splinters from explosions), brick walls designed against progressive collapse (to harden beams and pillars to withstand explosions), more control over gates (to prevent unauthorized access of persons and vehicles), fences and lighting (to detect and deter infiltrators), skylights with bars, door alarms, cameras with infrared detection, system redundancies, and backup power (Wade, 2002).

Contingency and Response Plans

After the vulnerability assessment and prioritization of resources to harden the facility, the manager should focus on the contingency and response plans in the event of a terrorist strike. Various kinds of emergency situations should be identified and coordinated plans developed. If there is an early-warning monitoring system in place, specific response plans must be prepared to deal with/prevent further contamination and to issue prompt advice to the public. Effective early-warning systems must be devised as networked systems including physical detectors of contaminants at the water source, treatment, distribution and storage facility, and extending to the medical care surveillance system to detect unusual occurrences or clusters of illnesses that may be linked to the water supply.

The emergency response plan must be developed with the participation of all major stakeholders. Stakeholders (Foran & Brosnan, 2002) can include the following:

- Individuals with specific expertise (e.g., microbiologists, toxicologists).
- Politicians/community leaders.
- Health department, hospital representatives, other healthcare professionals.
- Representatives of the local water utility.
- Representatives of water regulatory agencies (local, state, and federal).
- Representatives of high-risk groups.
- Staff from the wastewater treatment plant.
- Major water users and processors.
- Law enforcement agencies.
- Psychologists.
- Other emergency preparedness groups (e.g., fire department).
- Representatives of sources that pose potential threats to the drinking water system.

Exercises

Exercises are pre-planned events to test the alertness and preparedness of the staff and guards. They are also a way to inculcate a sense of vigilance among the staff so that they are constantly on the alert for any suspicious activities. Such exercises also hone the response capability of the staff in real emergencies. Exercises can be taken to the higher level of red-teaming, which are routinely employed by police and military forces. This involves the setting up of a special red-team the members of which wear the hat of terrorists.

The red team will survey the facility and probe for weaknesses. They will then try to infiltrate the defences physically and attempt to attack the facility by bringing in dummy bombs or gaining access to sensitive rooms. The staff involved in the exercise should be made aware that such exercises are not fault-finding missions to spot security lapses, but rather to identify areas for security improvement relating to staff processes and physical vulnerabilities. This will keep the staff vigilant at all times and also provide useful lessons the better to equip staff to manage real emergencies.

Conclusion

The intent of this paper is to urge international water experts to pause for a moment in their global water management debates to consider seriously the threat that terrorists pose to our precious water resource. By one stroke, a terrorist could wipe out years of hard work in building up a good water infrastructure.

It is the hope of this paper that there will be greater awareness among water practitioners of the role they can play in protecting current and future water facilities. It is a fact of life that mindsets and the status quo are hard to change. The paper harbours the optimism that with greater understanding of security threats to water systems, some real discussions might be generated; that at some point in the future, a water manager who is upgrading a water supply utility or a water expert who is planning a new water treatment facility, would ask the simple question: 'How might one protect this water facility from a terrorist attack?'

References

Asian Development Bank (2007) *Asian Water Development Outlook 2007: Achieving Water Security for Asia* (Manila: Asian Development Bank). Available at http://www.adb.org/Water/Knowledge-Center/awdo/default.asp/.

Biswas, A. K. & Tortajada, C. (2009) Changing global water management landscape, in: A. K. Biswas & C. Tortajada, C. (Eds) *Water Management in 2020 and Beyond*, pp. 1–34 (Berlin: Springer).

Burrows, W. B. & Renner, S. E. (1999) Biological warfare agents as threats to potable water, *Environmental Health Perspectives*, 107(12), pp. 975–984.

Courtney, E. J. (2003) Securing our water systems and their infrastructure, *Journal of Counterterrorism and Homeland Security International*, 9(3).

Doherty, B. (2002) Aqua-terror: don't drink the water, *Reason*, 33(10), p. 15.

Foran, J. A. & Brosnan, T. M. (2000) Early-warning systems for hazardous biological agents in potable water, *Environmental Health Perspectives*, 108(10), p. 993.

Gleick, P. H. (2000) *Water Conflict Chronology* (Online). Available at http://www.worldwater.org/conflict.htm (accessed on 1 July 2009).

Hamm, A. F. (2001) Water districts pump up security against bio-scares, *Silicon Valley/San Jose Business Journal*, 19(27), p. 53.

Khan, A. S., Swerdlow, D. L., Juranek, D. D. & Kahn, A. S. (2001) *Precautions Against Biological and Chemical Terrorism Directed at Food and Water Supplies*. Public Health Report (1974–), Vol. 116, No. 1 (Washington DC: Association of Schools of Public Health).

Lederer, E. M. (2004) UN calculates the cost of terrorism. *The Age*, 27 August. Available at http://www.theage.com.au/articles/2004/08/27/1093518081060.html?from=storylhs/.

Mader, B. (2003) Keeping the water safe, *Business Journal*, 21(4), p. A32.

Marta, L. (2006) Water security on tap, *Security Management Magazine*, 50(6), pp. 34–35.

Mays, L. W. (2004) *Water Supply Systems Security* (New York, NY: McGraw-Hill).

Meinhardt, P. L. (2005) Water and bioterrorism: preparing for the potential threat to U.S. water supplies and public health, *Annual Review of Public Health*, 26, pp. 213–237.

Oliver, R. (2007) *All About: Water and Health* (CNN), 18 December. Available at http://edition.cnn.com/2007/WORLD/asiapcf/12/17/eco.about.water (accessed 1 July 2009).

Petrasek, A. (2003) Water safety, *American City and County*, 1 June.

The Holy Bible, New International Version, Genesis 7:23.

United Nations Children's Fund (UNICEF) and World Health Organization (WHO) Joint Monitoring Programme for Water Supply and Sanitation (2008) *Progress in Drinking-water and Sanitation: Special Focus on Sanitation*. MDG Assessment Report No. 2008.

Wade, B. (2002) Locking down on system security, *American City and County Magazine*, January.

Zirschky, J. (1998) Environmental terrorism, *Journal (of the Water Pollution Control Federation)*, 69(7).

Singapore Water Management Policies and Practices

IVY ONG BEE LUAN

Lee Kuan Yew School of Public Policy, National University of Singapore, Singapore

ABSTRACT   *This paper explains Singapore's holistic approach and effective governance on water management in Singapore. Since gaining independence, an enabling environment which includes a strong political will has pushed the country successfully to achieve self-sufficiency in water. There are also legal/regulatory and institutional frameworks put in place to ensure effective implementation of its water management policies. The institutional framework has facilitated an integrated 'whole-of-government' approach to land-use planning, water management, a sound built environment and pollution control. Lastly, the technical framework that comes under its national water agency has effectively managed the entire water cycle as a single system for the whole country.*

Introduction

Water is a precious resource. Without it, you die. You can live without energy ... but without water you dehydrate and you die. The way water has been wasted around the world, misused, I foresee water shortages in many countries. Besides, earth warming [is] causing disruptions to the water supply and river basins and so on. I believe water reclamation and waste management will be a huge industry because almost every society, especially China and India, the big ones, will have to cope with this problem. ...

We did not do this by ourselves. We climbed on other people's shoulders. We brought this [technology] together and improved on them. We are happy to have people climb on our shoulders, whether you're from Middle East, China, India, whatever. It's a collaborative effort. The world will need this because what we have assumed was limitless—endless supplies of water—is not so. We have found it not to be so and we've got a way out of it.

(Minister Mentor Lee Kuan Yew, *The Straits Times*, 26 June 2008)

This paper explains the governance of water management in Singapore, focusing particularly on the institutional/legal framework that ensures effective implementation of its water management policies.

Singapore is a small city-state of about $700 \, km^2$, with limited natural water resources. It is densely populated with a population of 4.8 million. The country is located just off the southern tip of the Malay Peninsula and is linked to West Malaysia, its northern neighbour, by a bridge commonly known as the Causeway, which is 1.056 km long carrying a road, a railway, and water pipelines across the Straits of Johore (Motha & Yuen, 1999). Singapore's strategic position and its relatively small size has, throughout its history, given its population a heightened sense of the unique nature of the city-state, as well as its isolation and vulnerability.

Being situated in the equatorial rain belt, Singapore receives an average of 2,400 mm of rainfall annually, well above the global average of 1,050 mm (Long, 2002). The constraint faced by the country is its limited land area to catch and store the rainfall, and the absence of natural aquifers or groundwater. This resulted in Singapore being classified by the United Nations as a water-scarce country (United Nations Educational Scientific and Cultural Organization (UNESCO), 2006).

When Singapore became an independent state in 1965, its main source of water supply was water imported from Johor (Malaysia) and supplemented by water from the local catchments, i.e. reservoirs. The Singapore government recognized that these two sources of water would not be able to ensure a stable and sustainable water supply for the country's growing economy and population. The government thus had to explore other avenues for more sources of water supply.

Today, Singapore has widened its sources of water supply to come from four different sources known as the Four National Taps:

- Imported water (from Johor).
- Local catchment (reservoirs).
- NEWater[1] (recycled water).
- Desalinated water.

The Four National Taps provide Singapore with a diversified and sustainable supply of water through: (1) large-scale urban stormwater harvesting; (2) conducting large-scale recycling of water; and (3) desalination to augment imported raw water (Khoo, 2009).

Singapore's Approach to Water Policy

Singapore is able to achieve thus far in its water policy because of its strong governance that comes up with effective strategies. Singapore adopts a multi-pronged approach in its water supply and water-management policies, namely: physical infrastructure; legislation and enforcement; water pricing; public education; and research and technology.

Overarching all these is a holistic approach that encompasses the following frameworks, which is the main subject of this paper:

- Political.
- Institutional.
- Technical.

The institutional framework is given more attention in this paper as information on the other frameworks, namely political and technical, is readily available in the published literature.

Political

Political will. The Singapore government has been an essential force behind the successful water policy, its strategies, planning, and implementation. The long-term security of water was an important consideration for Singapore when it became a newly independent nation (Tortajada, 2006). It was in 1965 when Singapore was separated from Malaysia and became an independent state. It was the vision of Singapore's Minister Mentor Lee Kuan Yew, who was the country's first Prime Minister, that enabled the then Ministry of the Environment and the Public Utilities Board (PUB)[2] to achieve the Four National Taps Strategy.

On 9 August 1965, the then Malaysian Prime Minister Tunku Abdul Rahman had said that 'If Singapore's foreign policy is prejudicial to Malaysia's interests, we could always bring pressure to bear on them by threatening to turn off the water in Johor' (Lee, 1998). In 2008, Minister Mentor Lee shared that it was that comment and an earlier incident that drove home to him the need for Singapore to strive for self-sufficiency in water. The earlier incident was in February 1942 when the island fell to invading Japanese troops who blew up the pipes transporting water from Johor to Singapore. That left Singapore with only two reservoirs of water that could last at most two weeks (*The Straits Times*, 2008).

With that, MM Lee vowed with his engineers, 'from day one', systematically to try to turn every drop of water in Singapore into potable water. Since then, the 'quest for water independence' has dominated every facet of urban development in Singapore (*The Straits Times*, 2008).

Johore Water Agreements. The issue of supplying water to Singapore has, at times, been brought to the forefront of the Malaysian and Singaporean political arena (Lee, 2003). The most recent widely publicized political negotiations between the two countries on the water supply issue took place between 1998 and 2003. It resulted in the Singapore government taking the unprecedented step of releasing into the public domain in January 2003 correspondence between the two countries on the negotiations, as it was necessary to set the record straight and let people judge for themselves (Ministry of Information, Communications and the Arts (MICA), 2003).

The political subject of contention is the two water agreements signed between the Johor government (Malaysia) and Singapore's City Council (the predecessor of Singapore's Public Utilities Board) in 1961 and 1962, collectively known as the Johor Water Agreements. These agreements provide for Singapore to import raw water from Malaysia and, more specifically, for Singapore to draw up to 350 million gallons of water per day (mgd) (MICA, 2003):

(a) *Tebrau and Scudai Water Agreement*
   (i) Signed in 1961 and in force up to 2011.
   (ii) Grants PUB (then the City Council) the full and exclusive right to draw off and use water from Gunung Pulai Reservoirs as well as the Tebrau and Skudai Rivers.

(b) *Johor River Water Agreement*
   (i) Signed in 1962 and in force up to 2061.
   (ii) Grants PUB (then the City Council) the full and exclusive right to draw up to 250 mgd of water from the Johor River.

Both agreements were guaranteed by the Malaysian government in the Separation Agreement that established Singapore as a sovereign state in 1965, and were enacted into the Malaysian Constitution by an Act of Parliament (MICA, 2003). Under the Water Agreements, Singapore pays the Johor government 3 cents (RM0.03) for every 1,000 gallons (4,546 m^3) of raw water. In turn, the Johor government pays Singapore 50 cents (RM0.50) for every 1,000 gallons of treated water, although it costs Singapore RM2.40 to treat every 1,000 gallons of water. By selling treated water to Johor at only 50 cents (RM0.50), Singapore absorbs a subsidy of RM1.90 per 1000 gallons of water (MICA, 2003).

Both the Water Agreements contain a provision that allows for a review of water prices in 25 years' time, and arbitration in the event of a disagreement. Prices can be revised in line with the purchasing power of money, labour costs, and cost of power and materials used to supply water. In 1986 and 1987 (i.e., 25 years after 1961 and 1962, respectively), Malaysia did not revise the water rates because of financial considerations. The Johor government knew that if it were to raise the price of raw water, the price of treated water that Johor buys from Singapore would also go up (MICA, 2003).

In June 1988, the then Prime Minister Lee Kuan Yew and the Malaysian Prime Minister Dr Mahathir Mohamad signed a Memorandum of Understanding that provided for the sale of treated water, beyond the 250 mgd that Singapore is entitled to draw under the 1962 Water Agreement, under a new price formula. However, the price provided for raw water under the 1962 Water Agreement was not altered (MICA, 2003).

The situation that Singapore and Malaysia find themselves in over water is not unique (Kog, 2002). There are several countries sharing water resources with their neighbours. This is because their rivers and groundwater basins transcend national boundaries. Therefore, neighbouring countries of the water resource have a natural 'right' to the resource. Singapore's situation, however, is unique in that it is not a case of Malaysia sharing its water resources with Singapore, but a case of Singapore having to buy water from Malaysia, which owns the natural resources.

Water from local catchments and imported water from Johor are adequate for Singapore's needs (Goh, 2003). But the Singapore government decided to supplement them with NEWater and desalinated water. In this way, by 2011, when the 1961 Water Agreement expires, Singapore will not need to renew it. By 2061, when the 1962 Agreement expires, Singapore can be self-sufficient in water if there is no new water agreement with Malaysia (Tan, 2009).

Institutional

Singapore has to plan for many land uses within its small island. Besides housing, where it has managed to house its residents with 90% of them owning their own homes, land has to be set aside for uses such as commerce, industry, defence, waste disposal, water needs, as well as facilities to support its role as a major air hub and one of the largest container ports in the world (Cheong, 2008). All these are made possible because the government institutions have developed a good system of governance that adopts long-term comprehensive planning and have a steadfast commitment to sustainable development. These have enabled Singapore to achieve economic growth, a good standard of living, and a good environment in a balanced and pragmatic way (Cheong, 2008).

Land-use planning was first undertaken by the then Planning Department set up under the Planning Act in the 1960s (Motha & Yuen, 1999). The Planning Act provides for a Master Plan which is a statutory land-use plan that guides Singapore's development in the medium-term of 10–15 years.[3] In the 1970s, the Water Planning Unit was formed and the Water Master Plan was developed (Tan, 2009). This translated into integrated planning encompassing land use, environmental and water concerns, etc. across various government agencies. The various government ministries are clear about their roles and responsibilities, and work effectively together as a public sector in a coordinated way.

Putting in place good infrastructure.   In the area of water management, the government's strategy was first to ensure that the physical infrastructure and facilities were properly built in place to separate wastewater from clean water run-offs. Today, 100% of Singapore's population is scrved by modern sanitation.

The basic public physical infrastructural works were carried out by the public sector. Today, all types of premises and property developments, whether they are undertaken by the private sector or the public sector, are required to ensure that the premises/properties are properly constructed, maintained, and connected to the public sewers or proper sewerage facilities.

Inter-agency coordination.   Singapore has been very effective in integrating land-use planning and water management. This prevents water pollution at an early stage, and constitutes one of the most critical factors for successful catchment management. Effective cross-sector coordination among the relevant government agencies in water management ensures the success of integration, and reduces inter-sectoral conflict of interest (World Bank *et al.*, 2006).

There is a good institutional framework where the relevant government agencies effectively coordinate with one another for any application by both the private and the public sectors in development works.[4] Under the Simplified Planning Approval System introduced in April 1987 (Motha & Yuen, 1999), several government agencies are involved in the process that integrates the planning and technical requirements to ensure that all developments meet the legislative requirements on structural, environmental health, water supply and discharge, and occupational safety aspects. These agencies and their respective areas in-charge are shown in Table 1.

New developments.   Before a proposed development can be constructed under the Building Control Act, the developer must submit Building Plans (BPs) of the building works to the Building Plan and Management Division of the Building and Construction Authority (BCA) for approval.

Before BCA approves the BPs, they will have to be submitted and cleared by the various authorities, as listed in Table 1. The Central Building Plan Unit (CBPU) under the National Environment Agency (NEA), a 'one-stop' centre that coordinates response from the technical departments in PUB and NEA, issues clearance of BPs subject to compliance with sewerage, drainage, environmental health, and pollution control (on water, air and noise) requirements.[5]

These agencies ensure that the qualified persons, i.e. professional architects or engineers of any development, have consulted and obtained the necessary approvals from the relevant government agencies before their projects can be approved and the works commenced. For example, the PUB and NEA check on the sanitary and sewerage systems

Table 1. Government agencies involved in building development process

S/N	Government agencies to be consulted	In circumstances involving ...
1	Urban Redevelopment Authority (URA)[a]	Planning/land-use approval Buffer provision
2	Building & Construction Authority (BCA)[b]	Building control Building plans
3	Public Utilities Board (PUB)	Drainage provision Sewer layout Water supply Developments within water catchment areas
4	National Environment Agency (NEA)[c]	Siting of industrial/warehouse developments Environmental health and pollution control, including pollution to watercourses (for industrial premises) Refuse disposal requirements Developments within water catchment areas
5	Land Transport Authority (LTA)	Road proposals for the development including issues of access, planting of verges in car parks and parking provision
6	Fire Safety Bureau (FSB), Singapore Civil Defence Force	Fire safety Hazardous materials Dangerous trades
7	Singapore Power	Electricity loading
8	National Parks Board (NParks)	Public open space provision Tree felling Tree conservation areas

Notes: [a]URA is Singapore's national land-use planning authority. It prepares long-term strategic plans, as well as detailed local area plans, for physical development and then coordinates and guides efforts to bring these plans to reality. URA coordinates the efforts of various government agencies in implementing selected infrastructural facilities and key environmental improvement projects. Development control is the implementation arm of URA where it facilitates development by ensuring orderly and rational private sector development in accordance with its strategies and planning guidelines (see http://www.ura.gov.sg).
[b]BCA is an agency under the Ministry of National Development championing the development of an excellent built environment for Singapore. 'Built environment' refers to buildings, structures and infrastructure in the surroundings that provide the setting for the community's activities (see http://www.bca.gov.sg).
[c]NEA is an agency under the Ministry of the Environment and Water Resources. It is Singapore's leading public organization responsible for improving and sustaining a clean and green environment in Singapore (see http://www.nea.gov.sg).

in the submitted drawings to ensure there are proper piping systems and the proposed connections to the public sewers are in order.

Qualified persons have to comply with all the requirements under the Sewerage and Drainage Act (SDA), the Code of Practice on Sewerage and Sanitary Works, and the Sewer Protection Zone Requirements to avoid damaging the public sewerage system when carrying out piling and structure works. It is an offence (the penalty can be a fine of up to S$20,000) to carry out any building works or structure works above, over, and adjacent to any public sewers without the PUB's approval. In fact, the PUB has the power to require the person to demolish or remove the building or structure to protect the public sewerage system.

Before any earthworks and demolition are carried out, clearance from the PUB relating to whether there are existing sewers on the site and for clearance on the proposed platform levels must be first secured (Motha & Yuen, 1999).

Once the development/building works are completed, the BCA will ensure that the necessary clearances from the relevant agencies have been obtained before it issues a Temporary Occupation Permit (TOP) for the development and, subsequently, a Certificate of Statutory Completion (CSC).

With the above institutional mechanisms put in place, there are very tight controls and proposals for development are scrutinized to ensure they do not encroach on the public used water system (i.e. sewers, pumping, mains, etc.). This helps to avert any potential damage to the public-used water system and, in turn, prevents pollution resulting from overflow or leakage of used water (Tan, 2009).

Existing buildings—additions and alteration works. Previously, all additions and alteration works (A&A) to residential houses were required to be cleared by the CBPU before BP approval could be granted by the BCA. The practice was applied to all A&A cases, regardless of the scale and complexities of the A&A works involved. The principal purpose of this procedure is to safeguard any drainage reserves and public sewers that may be within the boundaries of the houses.

On 1 September 1994, the procedure was simplified by doing away with the CBPU's clearance for cases where drainage reserves and public sewers are not affected. However, the simplification does not mean a relaxation of the safeguarding of drainage reserves and public sewers. Instead, the qualified person has to declare on a prescribed form that is submitted to the BCA and CBPU that the proposed A&A works do not involve building extension over a public sewer or a new connection to a sewer; and does not encroach on any drainage reserve.

In summary, the institutional framework put in place in Singapore has enabled inter-agency coordination and support, thereby facilitating a 'whole-of-government' approach to ensure that all buildings/developments in Singapore are properly connected to the public water piping network (comprising the water supply system and the used water system), and do not encroach on the public-used water system (i.e. sewers, pumping, mains, etc.).

Legislations Relating to Water

Having put in place the essential infrastructure required to manage used water, legislations are enacted to ensure that the infrastructure is properly used and not tampered with. For example, stringent pipe-laying and sanitary-work requirements are imposed through legislation enacted by the PUB (Tan, 2009). The legislation also places an onus on industries and private owners who have to do their part in controlling water pollution and managing their used water.

A number of legislations have been enacted, with strict penalties to control the pollution of watercourses, to ensure that the various industrial activities do not have an adverse impact on water quality or the management of used water. These legislations help to ensure that:

- the physical infrastructure of sewerage and drainage systems in the island are always properly constructed, maintained, and connected to the main sewers, regardless of whether the property development is undertaken by the private or the public sectors;

- all wastewater (domestic and non-domestic) are discharged into the public sewerage system or proper sewerage facilities; and
- there is no pollution to the watercourses through improper discharge of trade effluent or hazardous waste (i.e. industrial water).

Protection of Public Sewers

Sewerage and Drainage Act (SDA). The SDA is the main legislation controlling sewerage infrastructure, and is administered and enforced by the PUB. The SDA is specific to used water, requires used water from all sources (domestic, industrial, agricultural and other premises) to be discharged into a public sewer if one is available, and to meet certain standards. Where a public sewer is not available, wastewater may, upon approval from the relevant authorities, be discharged into watercourses such as canals and drains upon meeting much more stringent standards (Economic and Social Commission for Asia and the Pacific (ESCAP), 2000). The Act also provides for penalties for offences that have resulted in water pollution.

The six Regulations[6] under this Act guide the provision of sanitary works in all premises, the provision of small sewage-treatment plants, the control of trade effluent discharged into the sewerage system, and the collection of fees to maintain the public sewerage system (Tan, 2009). Some of the key regulations are highlighted below.

Sewerage and Drainage (Trade Effluent) Regulations. This set of Regulations stipulates the limits of discharges of trade effluents into the public sewers and watercourses. Industries that discharge their wastewater into the public sewers are required to ensure that the quality of industrial wastewater complies with the stipulated limits. If not, they must install treatment plants at their factories to pre-treat their wastewater to within allowable limits before discharging into the sewers.

Under the Trade Effluent Tariff Scheme, applicants may discharge biodegradable used water with a higher concentration into the public sewer, subject to a fee.

In addition, industries can also dispose their organic sludge at designated water-reclamation plants for a fee. This provision offers a choice to industries that produce biodegradable wastes of higher concentration, but which find it undesirable or impossible to install, operate, and maintain a trade effluent-treatment plant on its premises (Tan, 2009).

Sewerage and Drainage (Sanitary Works) Regulations. These regulations explicitly stipulate the separation of rainwater from used water, and the diversion of rainwater into a surface stormwater drain and away from any opening connected to a used water system (Tan, 2009).

Furthermore, all sullage water from premises such as motor workshops, eating establishments, car washing bays in petrol stations, refuse chutes and bin centres, and backwash water from swimming pool filters are to be discharged directly into the used water system, via a grease trap where necessary.

Sullage water should be kept out of the rainwater-collection system as they contain pollutants such as detergents, organic material (food waste, oil and grease) and other harmful substances such as heavy metals from scrap metal yards, which will contaminate watercourses and catchments (Tan, 2009).

Sewerage and Drainage (Surface Water Drainage) Regulations. These regulations govern discharges into the stormwater drainage system by stipulating earth control measures/requirements for the proper discharge of surface run-off into the stormwater drainage system. They prohibit the discharge of silt or suspended solids into open drains in concentrations greater than 50 milligrams per litre (mg/l). The purpose is to protect the stormwater drainage system from silt and debris, which not only cause siltation and impede the effectiveness of the drainage system, but also contribute to the unsightly brown water in waterways and reservoirs (Tan, 2009).

The Regulations also require that every person carrying out earthworks or construction works should provide and maintain effective earth-control measures and take adequate measures to prevent 'any earth, top soil, cement, concrete, debris, or any other material to fall or be washed into the stormwater drainage system'.

Pollution Control and Public Health

Besides the SDA and its regulations that are administered by the PUB, there are two other Regulations administered and enforced by the NEA that are relevant to water and pollution to watercourses. These two Regulations are described below.

Environmental Protection and Management (Trade Effluent) Regulations. The Environmental Protection and Management Act deals with the control, in general, of environmental pollution. Under this Act, one of the regulations is the Environmental Protection & Management (Trade Effluent) Regulations in which the NEA regulates and sets standards of trade effluent discharged from trade/industrial premises into any watercourse. The NEA also regulates the discharge of toxic or hazardous substances into inland water by licensing trade or industrial premises that have trade effluent to discharge. The Regulations also provide for penalties or fines in case of infringements.

Environmental Public Health (Quality of Piped Drinking Water) Regulations. Under the Environmental Public Health (Quality of Piped Drinking Water) Regulations 2008, the NEA, as the public health authority, regulates and sets standards for the quality of piped drinking water. This is in exercise of the powers conferred by Sections 80 and 111 of the Environmental Public Health Act. The standards are based on the World Health Organization's (WHO) *Guidelines for Drinking-Water Quality*, 3rd Edn, Vol. 1, incorporating the First Addendum published in 2006.

The Regulation requires suppliers of piped drinking water to put in a water safety plan, and amongst other things, a monitoring plan. Although the PUB has its own Water Testing Laboratory that monitors the quality of drinking water to ensure it is safe and of high quality, this regulation administered by the NEA ensures there is a separate independent party that acts as a regulator and monitors treated water quality. Such independent checks by a separate regulating agency serve to reinforce Singapore's good water governance as there is greater transparency and accountability, and it leaves no room for piped drinking water standards to be compromised under any circumstances.

Summary of Legislations

Table 2 summarizes the key water-related legislations that ensure the proper use, maintenance, and management of the public sewers and drainage infrastructure that have been put in place. These are important as they form part of Singapore's single system of the water loop cycle.

Under the institutional framework, it is seen that legislations are, by themselves, not enough. The legislations must be implemented. In fact, legislations and a good institutional framework must go hand in hand to ensure the former is implemented effectively. In Singapore, the implementation of the various legislations relating to water is implemented not by one agency alone. Instead, they are implemented effectively by a few government agencies under a well coordinated institutional framework that provides an integrated 'whole-of-government' approach to land-use planning, water management, a sound built environment, and pollution control to the general environment.

Technical

Besides an enabling environment which includes a strong political will, effective legal and regulatory framework and institutional effectiveness, one main reason for Singapore's success in managing its water resources is its concurrent emphasis on supply and demand management, including wastewater and stormwater management (Tortajada, 2007).

This is the technical framework that encompasses the collection of water from the four sources, treatment of water, distribution of potable water, and the collection of used water, which then undergoes treatment or reclamation to produce recycled NEWater. The entire

Table 2. Main water-related legislations

S/N	Legislation	Purpose of legislation
1	Sewerage and Drainage Act	To provide for and regulate the construction, maintenance, and improvement of sewerage and land drainage systems To regulate the discharge of sewage and trade effluent into public sewers
2	Sewerage and Drainage (Trade Effluent) Regulations	Control the industrial discharge of wastewater and trade effluent into public sewers
3	Sewerage and Drainage (Sanitary Works) Regulations	Regulates sanitary works to ensure separation of rainwater from used water, and the diversion of rainwater into a surface stormwater drain. Also, regulates sanitary appliances to ensure proper sanitary plumbing and drainage systems
4	Sewerage and Drainage (Surface Water Drainage) Regulations	Regulates discharges into the stormwater drainage system through earth control measures and requirements
5	Environmental Protection and Management (Trade Effluent) Regulations	Control trade effluent discharged from trade/industrial premises into any watercourse Control discharge of toxic or hazardous substances into inland water
6	Environmental Public Health (Quality of Piped Drinking Water) Regulations	Regulates and sets standards for the quality of piped drinking water

water cycle is managed as a single system by the PUB, the national water agency overseeing all these functions.

Under the PUB Act, the PUB functions as an agent of the government in the management and maintenance of public sewerage systems, public sewers, and stormwater drainage systems, drains and drainage reserves belonging to the government. Accordingly, the PUB is vested with powers:

- to regulate the construction, maintenance, and improvement of sewerage and land drainage systems;
- to regulate the discharge of sewage and trade effluent; and
- to advise the government on all matters relating to the collection, production, and supply of water, and to sewerage and drainage.

Three operating departments in the PUB are responsible for ensuring the water supply (the production and supply of potable water, reclaimed NEWater and industrial water); water reclamation (the collection and treatment of used water); and the management/maintenance of the drainage systems. The PUB maintains an extensive network of some 990 km of drains and canals, and 7,000 km of public roadside drains. The drains and canals are regularly cleaned and maintained by contractors engaged by the PUB, so they are kept free-flowing without silt or debris (Tan, 2009).

A total of 100% of Singapore's population is now served by modern sanitation, and all wastewater is collected and treated. Used water is collected via a network of underground sewers linked to water reclamation plants. Instead of discharging all treated effluent to the sea, a portion of the treated effluent from the water reclamation plants is used to produce high-quality reclaimed water, which Singapore calls 'NEWater'. NEWater is made possible due to technological advancement that has enabled used water to be reclaimed after secondary treatment by means of advanced dual-membrane and ultraviolet technologies, at a reasonable cost. The excess treated effluent, which is not used, is then discharged into the sea. In addition, Singapore has a Deep Tunnel Sewerage System (DTSS) to channel used water to the water reclamation plants and then on to the NEWater factories (Ibrahim, 2005). The DTSS, conceived by the PUB, is an efficient and cost-effective solution for long-term used water management[7] to cater for Singapore's increasing population and expanding economy.[8]

With the fully sewered system that laid the groundwork for large-scale water recycling, and technological innovations to produce clean water from used water, the PUB's single-agency management of the entire water cycle as a single system has enabled Singapore to close its water loop.

Besides the technical aspects of the collection, treatment, and distribution of water, and the maintenance of the extensive drainage network, the PUB also engages in two very important aspects of water management: public education, and research and development.

Public Education

While the government through the PUB ensures that there is sufficient water to meet Singapore's needs, the PUB is making efforts to encourage Singapore to conserve water. The PUB recognizes the importance of individual efforts in water conservation and introduces demand management initiatives to promote this.

One of the most significant campaigns is the '10-Litre Challenge' launched by the PUB in 2006 to reduce domestic or household daily water consumption per capita by 10 litres.

The domestic sector accounts for 58% of the nation's total consumption. Through the outreach and educational programmes, the daily per-capita domestic water consumption reduced from 165 litres in 2003 to 156 litres in 2008. The PUB's target is to reduce this further to 155 litres by 2012, though this target might be reached sooner. On 27 April 2009, a new target to reduce domestic water consumption to 140 litres per person per day by 2030 was announced.

In 2008, the PUB launched the 10% Challenge to promote water conservation in the non-domestic sector, which accounts for 42% of the nation's total consumption. Under this programme, the non-domestic sector is challenged to save 10% of their monthly water consumption, thereby helping them save on their operation costs. The aim is to encourage users to manage water demand along the value chain, from sustainable and efficient design, to the use of water-conservation devices and water-demand management.

Some other schemes/initiatives include the Water Efficiency Labeling Scheme for water-related products. Under this scheme, distributors of water-related products such as taps, showerheads, dual-flush low-capacity flushing cisterns, urinals, and clothes washing machines were encouraged to label the water efficiency of their products. This allows consumers to make informed choices when making purchases. To enhance this scheme, the PUB implemented the Mandatory Water Efficiency Scheme on 1 July 2009 covering taps, dual-flush low-capacity flushing cisterns, and urinals and urinal flush valves.

The PUB's public education efforts and programmes are aimed at getting all users to conserve water. On water conservation, it is relevant to bring in the PUB's water-pricing policy. Singapore adopts a policy of charging for water and metering *all* customers. The water-pricing formula comprises the following components:

- Water Tariff.
- Water Conservation Tax—to reinforce the water-conservation message.
- Waterborne Fee and Sanitary Appliance Fee—statutory charges[9] payable to the PUB to offset the cost of treating used water, and the cost of the operation and maintenance of the public sewerage system is charged on the volume of water used.

The water pricing structure is such that domestic consumers face an increasing block tariff, while non-domestic and shipping consumers are charged a uniform volumetric rate. The increasing block tariff structure for domestic consumers is based on a volumetric component where a higher rate is charged if the household water consumption exceeds $40 \, m^3$ per month.

The domestic water price for the first consumption block is on par with the non-domestic rate, i.e. S\$1.17/m^3 per month. The next consumption block is priced higher so as to discourage a higher usage of water. Consumers will try to keep their consumption within the first block so they do not end up paying the higher rate. This policy is a reversal of the previous policy in the 1980s where domestic consumers were subsidized, by higher charges for industrial users, based on the notion that water was a social good (Khoo, forthcoming). While the previous subsidy policy resulted in domestic water consumption increasing steadily and outpacing that of population growth, the current pricing policy has been effective in water conservation, as seen from the declining domestic per-capita water consumption in the last few years.

Research and Development (R&D)

R&D is another of the PUB's strategy where it aims to increase water resources, keep costs competitive, and manage water quality and security. The PUB cooperates closely with its local and international partners to conduct applied R&D so as to help realize its mission of ensuring an efficient, adequate, and sustainable supply of water.

In 2006, the Singapore government identified water as a new growth sector and will invest about S$330 million in water R&D in the next five years. To develop further the water industry, the Environment and Water Industry Development Council (EWI) was set up to spearhead the development of the environment and water industry. The PUB supports the efforts of the EWI by facilitating water-related R&D work.

The Singapore government also has a few funding schemes to support environmental and water research. This allows the PUB to partner with companies, both local and international, to share the cost of performing R&D that can potentially lead to high-impact innovations and applications in the area of water.[10] In fact, Singapore's vibrant water industry has attracted reputable water R&D institutes and companies to set up water research and water-product manufacturing facilities in Singapore (Ibrahim, 2009).

The R&D does not stop with the PUB and its local and international partners. In June 2008, Singapore set up a new Institute of Water Policy (IWP) under the Lee Kuan Yew School of Public Policy, National University of Singapore, to perform research on water-policy and water-management issues, and to take on consultancy projects with governments and institutions such as The World Bank. The IWP will partner with the PUB to leverage on its sound water policy and governance, as well as operational and technical expertise in managing water resources in an integrated manner. The IWP aims to increase the profile of water issues in national policy agendas across Asia. In times to come, the IWP hopes to be the defining institution for leaders in the management of water issues. This will enhance Singapore's standing as a global hydro-hub, which is likely to see more developments and excitement in the years ahead.

Conclusion

Professor Asit K. Biswas, winner of the Stockholm Water Prize in 2006, observed that 'If the world faces a crisis it will not be due to physical scarcities of water, but it will be due to sheer mismanagement of water' (Biswas, 2006).

Indeed, given its vulnerability due to a lack of water resources and its small size, Singapore has overcome any hurdles and challenges to achieve what it has today— self-sufficiency in water. This paper has shown how Singapore, with a strong political will, good institutional, legal and technical frameworks that come with good governance, and good water management practices can transform a country faced with a life-threatening water-scarcity situation to a successful situation that offers opportunities for social benefit and economic gains.

Moving forward, Singapore is aware that it has to ensure that the overall water demand management programme will stretch its water resources to the fullest, while ensuring that water conservation efforts are never neglected. It also recognizes that it needs to develop further the water industry to become a dynamic part of its economy. Through continual R&D and collaboration with local and international partners, the country will strive to seek new ways to produce water at a cheaper cost and more advanced methods to augment its water supply.

The world is changing very rapidly, and with it the current water management practices must change as well (Biswas & Tortajada, 2009). For the past one to two decades alone, Singapore has made tremendous efforts to create comprehensive water supply and management systems. Technological advancement has made what previously seemed infeasible now a reality. Indeed, never before in human history has water management faced so many profound changes within a short period of time. The major challenge ahead for the water profession will be how to anticipate and manage these expected changes successfully and in a timely manner (Biswas & Tortajada, 2009). Given its vulnerability, Singapore will not be immune to these global changes. It is hoped with its various frameworks that have been put in place, coupled with its strong governance, that Singapore will continue constantly to look ahead, strive forward and meet the challenges of the future in its quest to ensure a sustainable good water supply for its people.

Notes

1. NEWater is a high-grade reclaimed water produced after treated used water has been further purified using a three-step process involving advanced membrane technologies such as microfiltration, reverse osmosis, and the final disinfection of the product water using ultraviolet light.
2. The PUB is a statutory board under the Ministry of the Environment and Water Resources. It is Singapore's national water authority and the agency that manages Singapore's water supply, water catchment and sewerage in an integrated way.
3. The Master Plan is translated from the Concept Plan, which is Singapore's long-term planning framework that maps out the vision for Singapore for the next 40–50 years in terms of strategic directions for land use and transportation. The Concept Plan is reviewed every decade to ensure it remains relevant. In the review, changing economic and population trends, and land-use needs are taken into consideration to guide Singapore's physical growth.
4. In Singapore, all matters pertaining to the development of land and the construction of buildings are the subject of statutory control in Singapore. Two main statutes form the basis of this control: the Planning Act and the Building Control Act (Motha & Yuen, 1999).
5. Code of Practice on Pollution Control (2000 Edn, with amendments).
6. The six Regulations that are enacted under the Sewerage & Drainage Act are: Sewerage and Drainage (Trade Effluent) Regulations; Sewerage and Drainage (Sanitary Works) Regulations; Sewerage and Drainage (Surface Water Drainage) Regulations; Sewerage and Drainage (Sewage Treatment Plants) Regulations; Sewerage and Drainage (Sanitary Appliances and Water Charges) Regulations; and Sewerage and Drainage (Composition of Offences) Regulations.
7. The DTSS is based on the simple concept that used water can be conveyed through deep tunnels using gravity to two centralized water reclamation plants. This would free up prime land, currently used to site the existing six water reclamation plants (WRPs) and 130 pumping stations, as well as the buffer land surrounding the WRPs.
8. As Singapore continues to grow and urbanize, it will need more pumping stations and expand water reclamation plants to collect and treat used water. These take up precious land and require huge investments in costly equipment. Thus, in 2000, the PUB started building the DTSS to cater to Singapore's increasing needs (see the PUB website at http://www.pub.gov.sg/dtss/Pages/default.aspx).
9. The Waterborne Fee and Sanitary Appliance Fee are statutory charges payable to the Public Utilities Board (PUB) under the Sewerage and Drainage (Sanitary Appliance and Water Charges) Regulations.
10. The funding schemes are: the Incentive for Research and Innovation Scheme (IRIS) by EWI; the Innovation Development Scheme (IDS) by the Economic Development Board; The Enterprise Challenge (TEC) by the Prime Minister's Office; and Innovation for Environmental Sustainability (IES) by the NEA (see the PUB website: http://www.pub.gov.sg/research/Collaboration_Opportunities/Pages/default.aspx).

References

Biswas, A. (2006) Water—managing a precious resource, *Pan IIT Technology Review Magazine*, 1(3).

Biswas, A. K. & Tortajada, C. (2009) Changing global water management landscape, in: A. K. Biswas, C. Tortajada & R. Izquierdo (Eds), *Water Management Beyond 2020* (Berlin: Springer).

Cheong, K. H. (2008) Chief Executive Officer, Urban Redevelopment Authority, Singapore, Panellist presentation on 'Strategies for Sustainable Urban Development in Singapore' given at the Forum on 'Sustainable Urbanization in the Information Age', United Nations Headquarters, New York, NY, USA, 23 April 2008. Available at http://www.ura.gov.sg/pr/text/2008/pr08-42.html/.

Code of Practice on Pollution Control (2000 Edn, with amendments in February 2001, June 2002, February 2004 and February 2009) (Singapore: Pollution Control Department, National Environment Agency). Available at http://app2.nea.gov.sg/data/cmsresource/20090312534898283541.pdf/.

Economic and Social Commission for Asia and the Pacific (ESCAP) (2000) *Wastewater Management Policies and Practices in Asia and the Pacific*. Water Resources Series No. 79 (New York, NY: United Nations).

Environmental Protection and Management Act, Cap 94A. Singapore Statutes. Available at http://statutes.agc.gov.sg/.

Environmental Public Health Act, Cap 95. Singapore Statutes. Available at http://statutes.agc.gov.sg/.

Goh, C. T. (2003) Then Prime Minister of Singapore, speech given at the official launch of NEWater at the NEWater Visitor Centre, 21 February 2003. Available at http://app.mfa.gov.sg/data/2006/press/water/SpeechPM.html/.

Ibrahim, Y. (2005) Minister of the Environment and Water Resources, speech given at the 'Light at the End of the Tunnel' Ceremony held to mark the completion of tunnelling works for DTSS at Mandai Road, 21 February 2005. Available at http://app.mewr.gov.sg/web/Contents/Contents.aspx?Yr=2005&Contld=570/.

Ibrahim, Y. (2009) Minister for the Environment and Water Resources, speech given at the Committee of Supply Debate 2009 at the Singapore Parliament, 9 February 2009. Available at http://app.mewr.gov.sg/web/Contents/Contents.aspx?Yr=2009&Contld=1272/.

Khoo, T. C. (2009) PUB Chief Executive, speech on 'Singapore—A Pioneer and Visionary City for Integrated Water Resource Management', given at the Water in the City of the Future, World Water Forum, Istanbul, Turkey, 20 March 2009. Available at http://www.pub.gov.sg/mpublications/Speeches/Speech21032009_1.pdf/.

Khoo, T. C. (forthcoming) *Water—Turning Scarcity into Opportunity*. Draft Paper, forthcoming.

Kog, Y. C. (2002) Natural resource management & environmental security in Southeast Asia—a case study of clean water supplies to Singapore, in *Beyond Vulnerability? Water in Singapore–Malaysia Relations*. Monograph No. 3 (Singapore: Institute of Defence & Strategic Studies (IDSS), Nanyang Technological University).

Lee, K. Y. (1998) *The Singapore Story—Memoirs of Lee Kuan Yew* (Singapore: Singapore Press Holdings).

Lee, P. O. (2003) *The Water Issue Between Singapore and Malaysia: No Solution in Sight?* Working Paper (Singapore: Institute of Southeast Asian Studies (ISEAS)).

Long, J. S. R. (2002) On the threshold of self-sufficiency: toward the desecuritisation of the water issue in Singapore–Malaysia relations, in: C. G. Kwa (Ed.) *Beyond Vulnerability? Water in Singapore–Malaysia Relations*, p. 109. (Singapore: Institute of Defence and Strategic Studies).

Ministry of Information, Communications and the Arts (MICA) (2003) *Water Talks? If Only It Could* (Singapore: Ministry of Information, Communications and the Arts).

Motha, P. & Yuen, B. K. P. (1999) *Singapore Real Property Guide* (Singapore: Singapore University Press).

Sewerage & Drainage Act, Cap 294. Singapore Statutes. Available at http://statutes.agc.gov.sg/.

Tan, Y. S. (2009) *Clean, Green and Blue—Singapore's Journey Towards Environmental and Water Sustainability* (Singapore: Institute of Southeast Asian Studies (ISEAS)).

The Straits Times (2008) Rising to the water challenge from Day 1, *The Straits Times*, 26 June.

Tortajada, C. (2006) Water management in Singapore, *Water Resources Development*, 22(2), pp. 227–240.

Tortajada, C. (2007) Water management in Singapore, Centre for Governance and Leadership, *Ethos*, 2.

United Nations Educational Scientific and Cultural Organization (UNESCO) (2006) *2nd United Nations World Water Development Report* (New York, NY: UNESCO).

World Bank, East Asia and Pacific Region, Environment and Social Development (2006) *Dealing with Water Scarcity in Singapore: Institutions, Strategies, and Enforcement*. Addressing Water Scarcity, Background Paper No. 4 (China: The World Bank Analytical and Advisor Assistance (AAA) Program) (Washington, DC: World Bank).

World Health Organization (WHO) (2006) *Guidelines for Drinking-Water Quality*, 3rd Edn (Geneva: WHO).

Appendix: Government Websites

Ministry of Environment and Water Resources
http://www.mewr.gov.sg
http://www/ifaq/gov.sg/mewr/apps/fcd_faqmin.aspx
http://app.mewr.gov.sg/web/Contents/ContentsEWI.aspx?Contld=431
http://app.mewr.gov.sg/web/Contents/Contents.aspx?Yr=2009&Contld=1309]

Public Utilities Board
http://www.pub.gov.sg
http://www.pub.gv.sg/general/Pages/SewerProtectionZone.aspx
http://www.pub.gov.sg/wels/about/Pages/default.aspx
http://www.pub.gov.sg/RESEARCH/Pages/default.aspx
http://www.pub.gov.sg/research/Collaboration_Opportunities/Pages/default.aspx

National Environment Agency: http://www.nea.gov.sg
Urban Redevelopment Authority: http://www.ura.gov.sg
Building & Construction Authority: http://www.bca.gov.sg
Attorney-General's Chambers: http://statutes.agc.gov.sg
Lee Kuan Yew School of Public Policy, NUS: http://www.spp.nus.edu.sg/Institute_of_Water_Policy.aspx

Water Management in Fiji

VINESH KUMAR

Ministry of Agriculture, Animal Health & Production Division, Suva, Fiji Istands

ABSTRACT *According to The World Bank, Fiji has one of highest per-capita fresh water resources in East Asia and the Pacific. However, these water resources are not evenly distributed—they are not equally plentiful in all places, nor is water equally available at all times. Above all, Fiji is an archipelago with a total of 332 islands (of which 110 are inhabited), hence managing water is a major challenge in itself. This paper tries to give a comprehensive outlook of water management in Fiji. It also outlines the key challenges for water management in Fiji and articulates broad recommendations. The paper concludes that the challenges of ensuring that water is conserved and managed wisely are huge and no single agency can address them in isolation. Strengthening partnerships among stakeholders (governments, the private sector, non-governmental organizations (NGOs) and donors agencies) is the way forward.*

Country Background

Location and Land Area

Fiji is a Melanesian island group located in the South Pacific at 175° E longitude and 18° S latitude (Figure 1). It is a group of islands or an archipelago with a total of 332 islands, of which only 110 are inhabited. The group has two large islands—Viti Levu and Vanua Levu—that have a higher population density and are hubs of economic activity. Comparatively many of the smaller islands are made of coral reefs and are low in elevation, thus unsuitable for habitation. The total land area of Fiji is 18,272 km^2. Topographically, Fiji is divided into three major classes: plains and valleys; low mountains and hills; and high mountains. These landforms are depositional—littoral or fluvial, erosional—fluvial erosion, mass movement or volcanic (Macfarlane, 2005).

Population

According to the latest population census of 2007, Fiji has a population of 837,271 people, that is equally distributed amongst the rural (49%) and urban (51%) populations. Unlike many developing and developed countries, Fiji has only 2.63% of its population above 70 years of age, whilst 83% of its population is below the age of 50 years.

According to the United Nations Development Programme *Report 2007/2008* (UNDP, 2007), in 2005 Fiji had a Human Development Index of 0.762 and was ranked 92nd of the

Figure 1. Map of Fiji islands.

177 countries. Furthermore, it was ranked 50th in terms of the Human Poverty Index, with a per-capita gross domestic product of US$6,049 (Figure 2).

Fiji lies in the oceanic tropical climatic zone. Hence, it has two major seasons: hot and wet, and cool and dry. The temperature during the hot and wet season, which falls between November and April, ranges from 26°C to 27°C. The cool and dry season is between May and October, and during this time the temperature ranges from 23°C to 25°C. Rainfall distribution is strongly influenced by the terrain of the islands because leeward sides of mountainous islands tend to be drier and windward sides tend to be wetter. On Viti Levu, for example, rainfall ranges from 3,000 to 5,000 mm on the windward side, and from 2,000 to 3,000 mm on the leeward side. Average rainfall across the country is around 2,000–3,000 mm per annum.

It is important to note that the 'wet season' replenishes water supplies for the subsequent 'dry season'. The wet season occasionally brings in tropical cyclones, hurricanes and typhoons; floods and landslides are common during these times. Fiji has also experienced some severe droughts. One occurred at the beginning of the 1986 dry season and extended through the 1986/87 wet season; another occurred in 1997.

Water

According to the World Bank, Environment Department (2004), Fiji has 34,690 m^3 per capita of fresh water resources, which is one of highest in East Asia and the Pacific. As compared with the year 2000, its per-capita total actual renewable water resources (TARWR) increased by 4% (AQUASTAT, cited by United Nations Children's Fund (UNICEF), 2006). Interestingly, according to the Mundi Index (Indexmundi, 2006), as of 1987, Fiji had a 28.6 km^3 of total renewable water resources, which was fairly low

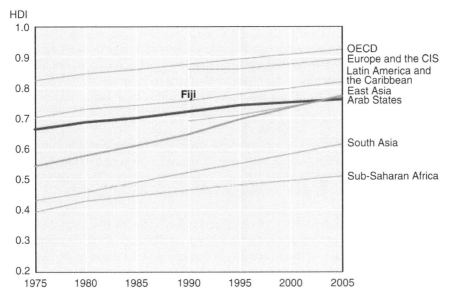

Figure 2. HDI-World indicator, position of Fiji (UNDP, 2007, Table 2).

as of that period. This entry provides the long-term average water availability for a country in cubic kilometres of precipitation, recharged ground water, and surface inflows from surrounding countries. Furthermore, it also stated that 'annual available resources can vary greatly due to short-term and long-term climatic and weather variations'. This gives an indication that there was a significant increase in the rechargeable ability of the water resources. Hence, it can safely be concluded that Fiji has sufficient renewable water resources for a while with the caveat that it is 'managed' well.

Variability in Water Resources

Though in total Fiji has a large water resource, in fact these water resources are not evenly distributed—they are not equally plentiful in all places, nor is water equally available at all times. The variability is very dependent on geographical location, variability in climate conditions, and socio-economic and environmental developments. Furthermore, a comprehensive study by Bronders & Lewis (1994), whereby they collected data during three years of fieldwork on small islands in Fiji, indicated that water resource problems, apart from climatological and geological constraints, are *mainly due to water-use practices*. The rainwater-harvesting system is a heavily under-utilized source of fresh water. The groundwater option is not an alternative, but a supplement to the existing water resources.

These water resources can be divided into two major categories: groundwater and surface water.

Groundwater

Groundwater in Fiji occurs on both the large islands as well as on small low-lying islands. However, its occurrence and challenges differ according to the different physical environments. 'Groundwater is found in superficial and medium-depth strata on the larger islands of Viti Levu and Vanua Levu and some large islands, in either fractured rock or sedimentary formations' (Pacific Islands Applied Geoscience Commission, 2007a). There are a number of groundwater aquifers across the country that vary in depth and volume. Currently, Fiji has ten water-bottling factories that bottle these groundwaters and export them. Though there is very little in the literature about the recharge ability of these aquifers, recently a joint statement by all ten producers claimed that their source was completely rechargeable:

> the water from our sources is a completely renewable resource that is constantly replenished by abundant rainfall each year. Moreover, the very livelihoods of our companies rely on the health and well being of our water sources. This ensures that the nation of Fiji will benefit from our sources for generations to come. (FijiLive, 2008)

On the other hand, groundwater plays a very integral part on the small islands. Many of the smaller islands have superficial groundwater lenses in sandbeds or coral formations, which lie on marine water. Many of these sources are constantly under threat as frequently they are not managed well. As stated previously, there is an uneven distribution of rainfall across the islands, and rainwater-harvesting using roof systems is widespread on these islands, but the psychology of rural people fails to take into account the possibility of extreme climate events and drought when there is relatively abundant water for most of the time (for instance, providing small-capacity storage instead of larger capacity). A few of the 110 smaller islands rely constantly on government vessels to transport water from the mainland. To make things worse, these vessels are often late and the people on these islands have to 'ration' their water supplies with other families. In very extreme cases, 'coconut water' can also be used as a substitute.

In many settlements around Fiji, groundwater is the major source for drinking water. The Ministry of Multi-ethnic and Provincial Development office over the years has provided assistance for many of these projects. In many instances, boreholes are dug that may run 60 to about 180 feet underground. Submergible electric water pumps are then used to draw water from the boreholes. However, it is significant to note that most of these boreholes are not regulated and in many instances the water quality is not inspected before the commissioning of such projects (Nuku, 2009).

Surface Water

Fiji, comparably with other smaller Pacific nations, has many rivers, creeks, a few lakes and some freshwater wetland. The majority of the urban water supply relies on surface water. Fiji also uses surface water for hydro-electricity generation. The largest hydro dam is the Monasavu hydro scheme, which was commissioned in 1983, consisting of four 20 MW generators. It also has two smaller ones with a capacity of 8.8 MW.

Surface water is also used by the agricultural sector, especially for irrigation and processing. In a few instances it has caused conflict based on priorities between direct consumers and agricultural users (Figures 3 and 4).

In general, given the rainfall and relatively intact forest cover that allows the capture and retention of water in underground aquifers, and the presence of several important perennial rivers and streams, the larger islands within the Fiji archipelago have adequate supplies of water to meet the needs of the population.

Figure 3. Surface water withdrawal by sector, 1987.

Figure 4. Surface water withdrawal by sector, 2000.

Water Institutions

Water institutions in Fiji can be categorized based on their usage and service delivery. The Public Works Department (Fiji Water Authority) looks after urban water supply and sanitation. Water used for irrigation is controlled by the Ministry of Agriculture. The City Council, Municipal Council, and the Public Works Department are institutions that look after urban drainage. The Ministry of Agriculture in limited areas puts in place measures to mitigate flooding. The Fiji Electricity Authority is the only agency that operates hydro dams to generate electricity (Figure 5).

Urban Water Supply

The urban water supply in Fiji has been under the jurisdiction of the Ministry of Public Works (Water and Energy), Works and Transport. Its core functions have been the provision of advice, technical services, planning, design and construction of works projects for other government departments and agencies. Furthermore, it provides for the management of works and maintenance programmes associated with water supplies and sewerage. However, in December 2006, the Ministry of Local Government, Urban Development and Public Utilities (MLGUDPU) took charge of the Water and Sanitation Department (WSD).

Legislative transition in the water sector. Currently, Fiji has an interim administration that is in the process of implementing radical changes in key governmental institutions through the 'People's Charter'. Under this, a Commercial Statutory Authority called the Fiji Water Authority (FWA) was proposed that was to replace the Water and Sewerage Department. However, following a successful lobby by non-governmental organizations (NGOs) and interest groups as well as based on the submission by the Minister for Public Enterprises and Public Sector Reform, the government shelved the programme (Fij-online, 2006). Instead, a new Fiji Water Act that replaced the Water Supply Act of 1950 was endorsed in July 2007, and a new Fiji Water Authority was established. Furthermore, during the same year, His Excellency, the President of the Republic of the Fiji Islands

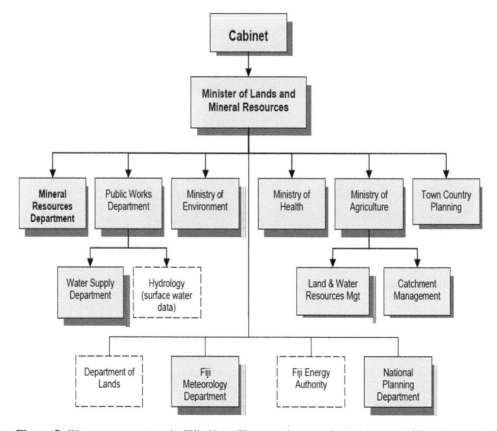

Figure 5. Water agency set up in Fiji. *Note*: The agencies or units in boxes not filled in are not formally members of the committee, but have important water-related roles.

established new functions and powers of the Water Authority of Fiji and its Board (South Pacific Applied Geoscience Commission (SOPAC), 2007).

Recent legal developments have been: (1) the drafting of new legislation to establish the Fiji Water Authority which will supply water to all towns in the country; and (2) draft amendments to the Minerals Act, which (a) establish a requirement to obtain a permit to extract groundwater (and to install bores and wells) within declared areas, and (b) limit polluting activities in declared areas, for the purpose of protecting the quality of groundwater. These drafts have been given cabinet approval to proceed, although they have not been through the parliamentary process.

Water Supply

Current Capacity of the Water Supply in Fiji

Currently, the WSD of Fiji operates and maintains 32 public water supply systems nationwide. The supply system is divided into two major categories: the city and town regional water supply that consists of 13 sub-systems, and the minor public system consisting of 19 subsystems. The whole water supply system of Fiji consists of 15 water-treatment plants and 110 service reservoirs and has over 2,200 km of underground water reticulation

Table 1. Water supply of Fiji, water connections and capacity

		Numbers of connections	Capacity (m³/day)
Central Eastern			
Kinoya	Urban	1,350	150,000
Debua	Urban	334	10,000
AdiCakabou	School		1,000
Wailada	Industrial		
Western			
Lautoka	Urban	5,200	45,000
Nadi	Urban	2,200	20,000
Sigatoka	Urban	120	4,000
Ba	Urban	250	6,500
Northern			
Labasa	Urban	780	6,000
Total		10,234	242,500

Source: Water Supply Department (2007).

pipelines. The sizes of these pipelines range from 50 to 900 mm nominal diameter. Table 1 provides information on some of the salient aspects of water supply in the country.

According to the WSD, currently it serves around 600,000 (nearly 75% of the population) people nationwide (Water Supply Department, 2007). However, this contradicts the World Health Organization (WHO)'s data that state that overall water supply coverage was only 47% (43% urban and 51% rural) as of 2004 (WHO, 2004). Practically, it is not possible for a 28% increase to have occurred over the last four years given the political and economic developments. Hence, the estimates provided by the WHO seem more credible.

As stated previously, the source of the supply varies from surface water to groundwater. The capital, which has the highest demand, extracts its raw water from Waimanu River, and in future it intends to extract it from Rewa River, one of the largest rivers in Fiji. A large part of Fiji's economy (approximately US$418 million annually in foreign exchange) is driven by the tourism sector. Nadi is the tourist town of Fiji that has many hotels and international resort chains. It is anticipated that developments in and around this area are growing at a significant pace. Currently, the major water source for these areas is a large dam at Vaturu, in the interior of Fiji, which was constructed in 1982. According to water experts, this water resource is not seen as a limiting factor, but in the present author's view it might become so if it is not managed and protected from contamination.

Most of the distribution systems in the towns and cities are designed to handle 150 litres of water per day per capita. However, most residents exceed this capacity, and on an average may consume from 200 to 500 litres per day per capita.

According to the Asian Development Bank (2003) 'the Water losses from leaking pipes or inaccurate or missing meters approached 55% of water supplied', and according to its latest report (Asian Development Bank, 2007), this has reached 70% in 2007. In past years the water supply sector has also been marred by scandals and inefficiency.

Water Losses in Fiji

Water loss occurs for the following reasons:

- Unmetered uses include fire-fighting and training; flushing water mains; sewers.
- Stolen unmetered water.
- Leaks in water mains.
- Leaks from hydrants.
- Leaks from valves (supply system).
- Water meter measurement errors in properties.
- Unmetered water tanks.
- Evaporation from uncovered reservoirs.
- Unmetered filling of swimming pools most often through fire hydrants.
- Reservoir overflows.
- Incorrect bulk meter readings.[1]

Reliability of the Service

Over the past two years the reliability of the water supply has reached 100% (especially in the capital and surrounding areas). However, in recent years power blackouts and pump breakdowns have caused some disruptions, especially in the greater capital areas. One concern is that the supply capacities of the storage tanks for these areas can only last for 12 hours. To mitigate the immediate effects, water is carted to these areas in portable tanks. Elsewhere the problem remains, as occasionally the media highlights disruptions in the supply due to collapse of the main supply line or 'burst' in the underground pipelines. It can take significant time and resources just to locate these leaks as they occur at a certain depth underground, let alone meet the cost of repairs. According to Deputy Permanent Secretary, Information, Major Leweni, he was 'privy to a study in 2005 on the water supply infrastructure which revealed that the infrastructure had been in place for almost 40 years'. Hence, time and again these problems are likely to arise, unless and until significant efforts are placed not only in extension, but also in repair and maintenance-upgrading of the existing systems.

Sewerage and Wastewater

The sewerage systems in Fiji are maintained by the Public Works Department in the towns and cities. Note that only certain parts of the water supply area are connected to the centralized waste treatment plants. For example, in the Suva-Nausori area (capital), which has the highest population density, only one-third of the population is connected to the centralized system, whilst others (270,000) are served by the septic tanks. The topography of these areas is rugged and the soil structure is impermeable, hence much effluent flows directly into streams and coastal water. Overflowing sewage from sewers—which in some localities are undersized and subject to blockages—and from poorly maintained sewage pumping stations also contribute significantly to water pollution (Asian Development Bank, 2002).

Urban and Rural Drainage

The respective Municipal councils are responsible for the urban drains in their towns and city boundaries. The major drains and stormwater outlets are maintained by the Public Works Department. In recent years, inefficient drainage system designs and the condition of these drains have been blamed for the frequent flash floods in the major urban centres. There are flaws in the design of many drains around the country, which lead to a slow discharge of stormwater away from these areas. Rapid and unplanned growth in many parts of the urban centres can be blamed for this. Many of these drains were designed during the Colonial era and have not kept pace with surrounding developments.

The Ministry of Agriculture maintains the rural and farm drains around Fiji. It works along with other stakeholders such as the Fiji Sugar Cooperation, sugarcane farmers, rice growers, vegetable and livestock farmers.

Department of Mineral Resources (in the Ministry of Lands and Mineral Resources)

This department has the authority for licensing the abstraction of groundwater to be used for the production of bottled mineral water. Fiji has large underground aquifers which are under pressure/threat from commercial operators-bottling companies. Currently, Fiji has ten water-bottling factories that bottle the groundwater and export it. Fiji is one of the largest producers of bottled mineral water in the Pacific. The brand 'Fiji Water' has made a huge impact in the United States market and currently claims to be the only water bottling company in the world to join the Carbon Disclosure Project Supply Chain Leadership Collaboration—to disclose fully the carbon footprint of its products. The department has policies in place that ensure that these sources are not over-exploited and remain sustainable. It also assists in safeguarding the aquifers from possible pollution and in the creation of buffer zones around them.

Constraints

The bottling companies over the years have grown to a scale whereby they have a huge impact on the economy and there seems to be a power shift whereby the industry has assumed 'total control'. An example is reflected in a case in July 2008 when Fiji's Cabinet approved a US$0.10/litre export duty on all mineral water exports and a same amount on excise duty on mineral water sold for domestic consumption. Mahendra Chaudry, the former Finance Minister, said the main reason for the new tax was to stimulate conservation of Fiji's natural resources. 'Mineral water is a scarce resource which will deplete and a fair share of returns has to be passed on to the nation', he said. However, the bottling companies staged a nationwide strike which forced the government to revert its decision and speculators also said that this also cost the then Finance Minister his job. Another interesting issue this case highlights is 'the regulator becoming regulated', which at times is 'dangerous'. Many commercial entities may manipulate regulations in their favour and lead to the ultimate destruction of the whole industry.

There is a need for the development of rules to limit extraction to sustainable levels as well as to create a buffer zone to protect the aquifers.

Water Pricing in Fiji

For domestic users, Fiji has a tri-band pricing mechanism. The first band is 60 lpcd, or $17 \, m^3$ per month. The second band is $33 \, m^3$ per month. The pricing details can be seen in Table 2. According to a study by the Asian Development Bank (2002), the present tariff bands are so wide that 60% of domestic consumption is satisfied in the first band rather than just basic needs, which should be interpreted at 60 lpcd, or $10 \, m^3$ per month. The

Table 2. Water and Sanitation Department (WSD) current water tariff structure

Category	Consumption bands m³/3 months	Consumption bands m³/month	Tariff (F$/m³)	Tariff[b] US$/m³[b]
Water				
(i) Domestic	1–50	1–17	0.153	0.072
	51–100	18–33	0.439	0.207
	Over 100	Over 33	0.838	0.396
(ii) Commercial	All units	All units	0.529	0.250
Sewerage				
(i) Domestic	All units	All units	0.200	0.095
(ii) Commercial	All units	All units	0.200[a]	0.095
Average tariffs (F$/m³)	*Water*	*Sewerage*	*Average*	
Domestic	0.284	0.200	0.484	0.229
Commercial	0.529	0.226	0.755	0.357
average	0.363	0.210	0.563	0.266

Notes: [a] The sewerage tariff is assessed for all major industries and is based on a tariff of F$0.200/m³ and average F$0.266/m³.
　　　[b] F$1.00 = US$0.45.
Source: Asian Development Bank (2002).

second band is too high to provide any effective demand-management role on average domestic consumption. Furthermore, it has also concluded that sewerage charges are far too low and cannot recover operations and maintenance costs.

Hence, the existing tariff structure fails to generate sufficient revenues even to meet the operations and maintenance costs. Though the current system favours the poor and average-income households, in the longer term it may compromise the availability and quality of the water supplied. This is directly as a result of inadequate funds to maintain the system.

Currently, meter reading is done quarterly (every three months), whilst billing is done monthly. This itself at times leads to other problems such as the inability of customers to pay a one-off large amount due to a large variation in the estimates. This leads to mutual adjustments and rebates that cannot be intuitively justified.

Currently, there seems to be a flaw in the existing institutional arrangements for collecting tariffs. The Water Rates Office collects tariffs and these funds are transferred to the consolidated account of the government not to the Water Supply Department's account. Hence, the water supply department has to rely for funds for operations and maintenance from annual budgetary allocations.

The following scenario was recorded based on the finding of a study by the Asian Development Bank (2002):

In 1998 Water Sales for WSD were estimated at F$19.4 million and collections at F$12.1 million. This represents a collection efficiency of 64%. Over the next three years while billings increased to F$23.1 million for 2001, collections were only F$13.2 million which includes arrears from previous years. Consequently at best the collection efficiency is now some 56% and accumulated arrears of outstanding bills is F$18.2 million.

Source of Surface Water Pollution/Contamination

Non-point source. Fiji has a fair amount of industrial and agricultural activity. Most widespread of all is the sugar industry. A wide range of pesticides and weedkillers are used by this industry. Many sugarcane farmers (about 21,000 in total) are illiterate or unaware of the possible contamination of the waterways because of their negligence. In many instances farmers normally wash their knapsack sprayers and fertilizer bags directly into the waterways. Poor cultivation methods and constant land clearing triggers soil erosion and excessive silting in the waterways. Vegetables and other cash crops are also grown in the lower plains and along the major riverbanks in Fiji which are constantly subjected to flooding. Hence, chemicals and fertilizers constantly get washed into these natural waterways.

According to the *GEF Hotspot Analysis Diagnostic Report, Fiji Islands* (Pacific Islands Applied Geoscience Commission (SOPAC), 2007b), other non-point sources for contamination of surface water are as follows:

- Soil erosion resulting from exposure of the soil, leading to increased sediment discharges, high turbidity and colour problems due to extensive or inappropriate clearing of the native forest as part of logging operations or for agriculture; poorly designed or constructed unsealed roads and unplanned development activities and fire are used to clear undesired weeds in farming and forestry areas.
- The erosional effects of tropical forest clearing for agriculture and urbanization, of road construction and other activities in surface water catchments. A steep island topography causes floods, landslides, and sometimes major losses of vegetation and significant soil erosion.
- Runoff from agricultural land containing nutrients (from fertilizers) and sometimes toxic agro-chemicals (pesticides and herbicides).

Point source. It is interesting to note that all four sugar mills in Fiji (Lautoka, Rarawai, Penang and Labasa) are located near waterways. Since there are flaws in the environmental regulations and enforcement, it is anticipated that all the mills perhaps discharge a significant amount of their water pollutants into these waterways. One of the mills located in the northern part of Fiji known as Labasa has come under much scrutiny as a large waterway beside the mill has turned into a 'deadzone' (the waterway has no marine life and has a very foul smell—it is basically dead).

Fiji also has a gold-mining industry. The largest mine is located in the district of Tavua. 'Notoriously', it is known to discharge harmful chemicals into Nasivi Creek—notably cyanide. There have been reported cases of livestock deaths due to chemical poisoning in the lower ends of the creek. It could be for economic reasons that tough measures are not taken against the mining industry.

Other industries such as motor servicing, the battery industry, and oil companies are also a major threat to Fiji's water sources.

A survey to identify and quantify the volume of unwanted persistent organic pollutants (POPs) and associated contaminations in Fiji found that stockpiles of pesticides are the major environmental threat.

Another source of water pollution in Fiji is microbial pollutants. This is generally due to unsafe disposal (untreated waste water), vector control, and solid waste management in Fiji. These discharges occur from outfalls (point-source pollution) and from more diffuse flows from on-site sanitation systems within urban areas of surface water catchments. The rapid urbanization process is putting great pressure on both surface water (and groundwater) supply catchments used for urban and nearby rural water supplies. A study of water quality in the Ba River (Pacific Islands Applied Geoscience Commission (SOPAC), 2007b) and estuary found them to be seriously contaminated, with the dominant source of faecal contamination being Ba Town. Other sources of microbial pollutants direct faecal contamination of catchments and streams from animals (e.g. cattle, pigs). Commercial livestock farming, especially the dairy farms and pig farms, usually dump their waste into the waterways. Another common source are the improvised pigpens located along the riverbanks or on the coasts around the country.

The Ministry of Environment is responsible for the regulation of the point-source pollution, whilst the Ministry of Agriculture is deemed responsible for the non-source pollution. Though both Ministries have put some effort and measures to contain the problem, it needs a collaborative effort from other sectors and the general public. A lack of funds for enforcement and litigation are some of the key constraints. Public awareness and creation of endowment amongst the resource owners could alleviate this problem to a great extent.

According to a draft national water policy formulated by the National Water Committee of Fiji that was approved by Cabinet (December 2005) as an interim policy, the following policy measures need to be adopted in order to protect the water quality in Fiji:

- Water sources of good quality must be protected from depletion and pollution by adequate protection mechanisms, consistent with the rights and interests of those who may be affected.
- The control of point sources of pollution of water must be applied comprehensively to ensure that receiving waters in Fiji are protected from all artificial discharges, including sewage and discharges from industry and mining.
- Measures for controlling water quality degradation from non-point sources, such as soil erosion and catchment activities, need to be strengthened.
- Small-scale and fragile water sources (such as shallow island aquifers) which have value for drinking and domestic use should receive particular attention for their protection.
- The impact of rivers on the quality of coastal waters should be recognized and investigated and, where necessary, measures taken to prevent coastal degradation.

Water Quality Monitoring in Fiji

Fiji does not have specific water quality standards. However, it currently uses the WHO Drinking Water Quality Guidelines. Urban water quality in Fiji is monitored by the Public Works Department's (now WSD), National Water Quality Laboratory, located at the Kinoya waste treatment plant. Another government agency that assists in urban water quality monitoring is the Public Health Department of Ministry of Health. The Ministry of Health also tests water quality from the private and the rural sectors. The Institute of Applied Sciences, which is affiliated to the University of South Pacific (USP), also randomly samples urban and rural water supplies. Its services are also utilized by the private bottling companies and hotels. Another intergovernmental organization, the South Pacific Applied Geoscience Commission (SOPAC), is also engaged in research and water hygiene management in Fiji. It has been shown that higher fluoride levels occur in the Suva water supply system than elsewhere around the country.

Some Cases of Water Borne Diseases in Fiji

On 1 July 2005, Radio New Zealand reported that 'Fiji's Public Works Department has admitted that its water supply system could be responsible for more than 10,000 cases of stomach ailments in the country.' Furthermore, according to a Fiji television report, cases of gastroenteritis estimated to have occurred between 1995 and 2000 were at a rate of over 2,000 a year, or nearly 170 a month.

The Fiji Times (11 February 2006) reported that in the Labasa area the number of cases of influenza and diarrhoea increased sharply, believed by health officials to be a result of flooding around the country. One farmer was reported to be in a serious condition with leptospirosis, a waterborne disease caused by the urine of livestock contaminating water; and there have been reports of typhoid.

According to a report by Kingston (2004), the two government departments (WSD and Ministry of Health) that carry out sampling and quality tests do not share information or data. This lack of synergy may lead to task duplication; however, in contrast, the practice will assist the departments to compare the consistency of their results.

Water Management Challenges in General

Hence, from the above, it can be concluded that the challenges in water management in Fiji are as follows:

- Increasing pressure on water resources due to upgrading and expansion of water sewerage systems around the country.
- Growing industrial, energy, mining, and commercial developments that demand more water.
- Increasing threats to water quality due to:
  ○ increased urbanization;
  ○ intensified agriculture, forestry and exploitation of natural resources; and
  ○ improper waste disposal.

Future Direction

National Level Collaboration

Firstly, there should be a collaborative effort from all sectors in recognizing the fact that surface water and ground water are critical for human well-being which will also provide growth for the country and will maintain the value of the natural environment.

Secondly, everyone (those in the commercial sector down to the grassroots population) has also to realize the fact that water resources are finite and these sources can be exhausted if not managed efficiently.

Thirdly, the effective management of water must be on the basis of the hydrologic unit—meaning, for surface water, the catchments of streams and rivers and for groundwater the aquifer system. Only in this way will the impacts of water exploitation in any location be adequately recognized and attended to. Surface water and groundwater are part of a unified water cycle and should be managed consistently as elements of the water resources of Fiji.

Water Conservation and 'Right' Pricing

One way one can conserve water in Fiji is by pricing it 'right'. As stated previously, most of the funding for water and wastewater comes from the revenues generated by pricing. Current pricing is not even enough to cover operations and maintenance, let alone capital expenditure. Therefore, Fiji must have water pricing that accurately reflects the true costs of providing high-quality water and wastewater services to consumers both to maintain infrastructure and to encourage conservation.

> Increasing water supply to meet continually higher demands generally has been the case in the past, is no longer a viable option for the future because of economic, social and environmental constraints, as well as physical availability of water. Water pricing has to be a part of the overall solution in balancing demand and supply. (Biswas, 2007, p. 222)

According to the European Union, the main and important goals of European Union water policy are the protection and improvement of the aquatic environment and the contribution to sustainable, balanced, and equitable water use. Water pricing is one in a series of possible tools to help achieve these goals (Roth, 2001).

Another school of thought is by environmental economists who argue that the more one draws from the nature, the less is available for future generations. The sustainability criterion suggests that, at a minimum, an allocation must leave future generations no worse off than current generations. Environmental economists have long advocated bringing the price mechanism more fully in line with 'full costs' so that 'users' might respond to 'market signals'—reflecting the true and full costs of production and consumption. Since water is basic to life, and certainly to one's quality of life, the pricing of water can be a powerful means of signalling this importance and scarcity to water users.

Another hard-to-digest fact is that nothing is free in this world. In fact, what seems free is being paid for by someone somewhere. It is an accepted attitude in developing countries that people 'don't care, it comes free'. People do not utilize resources wisely and efficiently if they do not have to pay for them.

Water Quality Protection

Fiji has an abundance of fresh surface water and groundwater sources and the increasing rate of rechargeability suggests that there is less probability of a physical shortage of water. However, when one looks at both the point and the non-point sources of pollution in Fiji, it is more likely that the shortage would be due to pollution, or in other words, shortage of clean water. But the challenge of protecting the water source is difficult as these are a number of polluters and it is not easy to isolate one from another. Hence, a collaborative effort is required from the whole community.

The Pacific Islands Applied Geoscience Commission (SOPAC) (2006) outlines the following principle, which is in line with the protection of water quality:

- Water sources of good quality must be protected from depletion and pollution by adequate protection mechanisms, consistent with the rights and interests of those who may be affected.
- The control of point sources of pollution of water must be applied comprehensively to ensure that receiving waters in Fiji are protected from all artificial discharges, including sewage and discharges from industry and mining.
- Measures for controlling water-quality degradation from non-point sources, such as soil erosion and catchment activities, need to be strengthened.
- Small-scale and fragile water sources (such as shallow island aquifers) which have value for drinking and domestic use should receive particular attention for their protection.
- The impact of rivers on the quality of coastal waters should be recognized and investigated and, where necessary, measures taken to prevent coastal degradation.

Institutional Strengthening

There are a number of water agencies in Fiji that normally work in isolation or with minimum collaboration. It has to be noted that most problems in this era are classed as 'complex problems' that touch upon several arenas simultaneously and require governmental responses that involve multiple jurisdictions and departments for effective resolution. Hence, the water agencies need to strengthen the coordination arrangements between water-use sectors and the various administrative units that deal with those water resources. This relation or inter-agency will create synergy and maximize resource utilization.

Stakeholder Involvement and Commitment

The challenges of ensuring that water is conserved and managed wisely are huge and no single agency can address them in isolation. Strengthening partnerships among stakeholders (governments, the private sector, NGOs, and donors agencies) is crucial for any policy implementation. Such cooperation can be factored into the action agendas and stakeholders' partnership agreements, which can be established to foster a sense of commitment and responsibility into any community-awareness programmes undertaken by government. These partnerships can complement each other and in many instances pool scarce resources for a common goal. These collaborations can be at country, regional, and global levels.

Conclusion

Water management in Fiji is a complex issue that permeates other sectors and cannot be easily solved. Hence, a many-sided approach is the only way to improvement. Education, training, investment in existing water resources, and safeguarding the groundwater resource for future use are potential solutions to the problem. Not only the water agencies with their special and vested interest, but also an integrated community approach are the way forward for the woes of water management in Fiji.

Note

1. Adopted from Water Demand Management Workshop Nadi, Fiji Islands, Skylodge Hotel, June 1999.

References

Asian Development Bank (2002) Fiji case study—Suva-Nausori, in: *Water Supply Fiji Islands*, pp. 11–16 (Suva: Asian Development Bank).

Asian Development Bank (2003) *Loan to Help Deliver Better Water and Sewerage Services in Fiji*, 18 December (Manila: ADB).

Asian Development Bank (2007) *Asian Water Development Outlook* (Manila: Asian Development Bank).

Biswas, A. K. (2007) Water as human right in the MENA region: challenges and opportunities, *International Journal of Water Resources Development*, 23(2), pp. 209–225.

Bronders, J. & Lewis, J. (1994) Water resources problems on small islands in Fiji. Paper presented at the 'Water Down Under 94' Conference, Groundwater/Surface Hydrology Common Interest Paper, pp. 405–409 (Adelaide: National Conference Publication—Institution of Engineers).

Fiji Times (2006) Water borne diseases on the rise. *Fiji Times*, 11 February, p. 4.

FijiLive (2008) *Fiji Water Bottlers Stop Production, Protest Over New Export Duty*. July. Available at http://www.fiji.gov.fj (accessed 3 March 2009).

Fij-online (2006) Available at http://www.fiji.gov.fj (accessed March 2009).

Indexmundi (2006) *Total New Able Water Resources—Fiji*. Available at http://www.indexmundi.com/fiji (accessed March 2009).

Kingston, P.A. (2004) *Surveillance of Drinking Water Quality in the Pacific Islands*. Situation Analysis and Needs Assessment Country Reports (WHO: Geneva).

Macfarlane, D. C. (2005) *Country Pasture Profiles. 2005*. Available at http://www.fao.org (accessed March 2009).

Nuku, W. (2009) Interview by Vinesh Kumar, District Officer, Tavua/Nadarivatu (13 March).

Pacific Islands Applied Geoscience Commission (SOPAC) (2006) *PfWG Fiji Report of Initial Mission* Discussion Paper SOPAC Programme for Water Governance (Fiji) (Suva: SOPAC).

Pacific Islands Applied Geoscience Commission (SOPAC) (2007a) Available at http://www.pacificwater.org (accessed March 2009).

Pacific Islands Applied Geoscience Commission (SOPAC) (2007b) *GEF Hotspot Analysis Diagnostic Report, Fiji Islands* (Suva: SOPAC).

Roth, E. (2001) *Water Pricing in the EU: A Review*. Euro-Mediterranean Information System on Know-How in the Water Sector International Portal. Available at http://www.emwis.net/topics/waterpricing/water-pricing-eu-review (accessed 9 February 2009).

UNDP (2007) *UNDP Human Development Report 2007/2008*. Available at: http://hdr.undp.org/en/reports/global/hdr2007-2008 (accessed March 2009).

United Nations Children's Fund (UNICEF) (2006) *UN World Water Development Report No. 2, 2006* (UNICEF).

Water Supply Department (2007) *Annual Report* (Suva: Water Supply Department).

World Bank, Environment Department (2004) *Environment at a Glance 2004, Fiji*. November. Available at http://siteresources.worldbank.org (accessed March 2009).

World Health Organization (2004) *The World Health Report 2004, Fiji*. Available at: http://www.who.int/whr/2004/annex/country/fiji/en/.

Urban Water Systems—Factors for Successful Change?

PONG KOK TIAN

Lee Kuan Yew School of Public Policy, National University of Singapore, Singapore

ABSTRACT   *What makes urban water system reform a success in some places, but a failure in others? Is there a fixed formula for success? This paper does not pretend to have the answers to the above questions. Through a focused study of the urban water system management experiences of Santiago, Chile, between 1990 and 1998 and that of Singapore, and by drawing parallels between the two, this paper attempts to identify, despite the different historical, political, social and economic context, some common factors behind their successful urban water system reform.*

Introduction

This paper begins by examining the circumstances surrounding the reform of Chile's Santiago Metropolitan Sanitary Works Enterprise (Empresa Metropolitana de Obras Sanitarias, or EMOS) in 1990, the key characteristics that made the reform a success, and the performance of EMOS post-reform till 1998. It will also study another urban water system management success – that in Singapore. By drawing parallels between the two, the paper attempts to distil out the common conditions/factors that made urban water reform possible.

The paper draws materials from the work of Shirley *et al.* (2000), which examined the reform of EMOS from 1990 to 1996. This portion is a focused and time-bound study of Santiago's urban water system, not an extension of on-going debate among water experts about the effectiveness and efficiency of Chile's unique Water Code in facilitating a water rights market. For the Singapore case, this paper draws material from the work of Tortajada (2006), which analyses the key reasons behind the country's success in water management. While drawing parallels between the two cases, this paper will also draw materials from the work of Tan *et al.* (2009).

Santiago's Urban Water System Reform

Political Context

An understanding of the factors behind EMOS's reform cannot be fully established without first knowing the political context of its reform. When EMOS was reformed in the late 1980s under public ownership, Shirley *et al.* (2000) saw none of the usual reasons to provoke it. First, there was clean, abundant, and cheap raw water supply from the Maipo River. Second, there was no ground pressure for EMOS reform as demand for water and sewerage in Santiago were largely met. Among EMOS's potential clients, 98% had water connections, while 90% had sewerage connections.[1] Third, EMOS's management was efficient and effective by regional standards. The extent of metering and bill collection rate were high (approximately 100% and 80% of 4.7 million connections), unaccounted-for-water was low (31% of total production in 1989), while staff/1000 water connections was reduced from 2.4 in the early 1980s to 2.1 in 1990. While finance was an area where EMOS had some problems, it was insufficient reason to precipitate reform. Between 1980 and 1987, EMOS ran a loss before taxes and was not allowed to increase its tariff (in real terms). Not able to invest more in network maintenance and facility upgrade, it suffered 52 breaks/100 km of network, with one-fifth of its pipes exceeding their 30-year usable lifespan. However, Shirlet *et al.* found its rate of return still better than all of Chile's public water companies, and its network and overall service was good by regional standards.

Shirley *et al.* traced EMOS's reform to the free-market-oriented Augusto Pinochet Administration (that came to power in 1973). Although the Administration had decided early on to sell EMOS, the eventual move to privatize EMOS was delayed for years. First, the small scale of water problems in Santiago rendered EMOS low on Pinochet's priority list. Second, many potential beneficiaries of a reform of EMOS, from the lower and middle class, were not considered important constituencies for Pinochet. Third, privatization in other sectors (telecommunications and electricity) with higher returns than in water gained priority. Fourth, Pinochet, confident of his political longevity, thought he had a longer window of opportunity to privatize water. The lack of urgency notwithstanding, within the larger context of Pinochet's interest to improve the regulatory framework and the performance of severely under-performing water utilities countrywide (especially those in his constituency), the sale of EMOS remained on his radar. Pinochet was eventually to lose the 1988 plebiscite and the military-backed candidate of the 1989 national elections, but the Administration still managed to pass new water legislation for EMOS's privatization before Patricio Aylwin's 17-party coalition came into power in 1990. The new legislation was practically irreversible as under the 1987 Constitution the government needs an absolute majority in Congress before it could change any legislation. The binominal electoral system (drawn up by Pinochet Administration after the 1988 plebiscite) systematically over-represented the military-backed coalition, making it almost impossible for the new Administration to control such a majority.

Aylwin's Administration was caught in a bind as it had won on the electoral platform that state enterprises providing basic services should 'stay in public hands'. His Christian Democrats had even pledged that privatizations after the 1988 plebiscite would be reversed. To their credit, they took a pragmatic approach and went around the problem by designing a reform meant for a private firm but implemented in state-owned enterprise. Their considerations were several-fold. First, there were net political benefits since the urban poor, a large constituency of the Administration, would benefit the most from more connections. Second, they could still justify their political platform, one of which was to retain the earlier regime's market-oriented and 'fiscally responsible' policies, and combine these with increased social spending. Third, reform of EMOS would be necessary to meet the anticipated increase in connection demand when the new Administration provided more housing for the urban poor. Fourth, expanded connection coverage could potentially increase government revenue through increased tariff collection. This could help fund expansion in social spending that the Administration had committed to, and also help the Administration side-step Constitutional constraints on government spending and flow of funds among Ministries. Pragmatic political considerations were key to bringing about the EMOS reform, which would not have occurred with no other precipitating factors of reforms.

Key Areas of Reform

Shirley *et al.* identified three key areas that the reform of EMOS had addressed effectively, whereas similar efforts in other countries had failed in the past:

- Government's information asymmetry with regard to management of water firms. Post-reform, EMOS and other water companies were required by the government to produce audited annual accounts according to generally accepted standards. The government created, based on legislation, an independent and professional regulatory body called the Superintendency of Sanitary Services (SSS), sending a credible commitment signal about the reform process and protection of consumers' benefits. The SSS was a small exclusive outfit comprising professional staff (all graduates) paid above-average civil service salaries, ensuring recruitment and retention of good staff, while reducing the risk of market capture. The SSS focused on core functions of a regulator, ensuring that EMOS complied with technical standards and investment plans. It also measured costs and efficiency from tariff adjustments, which were carried out every five years as legislated. Shirley *et al.* found that the main improvement was in accounting information, whereby failure to submit regular information on costs and service to the SSS could land the water companies with a fine. This reduced the companies' ability to manipulate their calculations.
- Incentives for the firm to comply with contract. In the past, tariff decisions were subject to much political discretion, with increases happening only when it was politically untenable to ignore unmet demand and deteriorating services. Post-reform, tariffs were higher, with greater transparency, consistency and public accountability. The regulatory framework, which was legislated, defined clearly the formulae and variables to be used for tariff-setting, with details even on calculating the peak and lull monthly tariffs. From 1989 to 1995, water and sewerage prices increased from US\$0.11 and US\$0.04/m^3 to US\$0.28 and US\$0.11/m^3, respectively. The five-yearly tariff adjustment was based on recovery of the long run average costs of a 'benchmark company', with a 7% return on capital. The construct of the 'benchmark company' was a black box, which prevents gaming. Water firms, including EMOS, were incentivized to be more efficient than the 'benchmark company' to earn additional profits. To balance the benefits to consumers, the regulations provided for the tariffs to be adjusted downwards at the end of the five-year period to force the company to share its gains with consumers. Private investors, therefore, felt assured that their return on capital would not be eroded through under-pricing, while the government showed its commitment to reduce the risk of monopoly rent. Shirley *et al.* highlighted that supportive institutions, such as a professional and

honest civil service, had helped ensure the incentive effects of the new tariff policy were not undermined by any residual information asymmetry and usual agency problems.[2]

- Credible commitment by the government to enforce contracts and adhere to promises. Chilean authorities demonstrated a credible commitment to honour its promises through regulatory contracts that specified 'neutral and automatic' enforcement. It allowed for water companies to appeal a dispute over tariffs to the SSS within 30 days. Failure to reach an agreement between the SSS and the company concerned would activate an arbitration mechanism, whereby a three-member arbitration panel would reach a decision within 37 days that both sides would have to accept. As a further recourse, the panel decisions could also be appealed to the courts. Shirley et al. highlighted the successful appeal in 1995 by EMOS for a higher tariff rate than was initially adjusted by SSS. (Bauer (1998), however, pointed out that Chilean courts tended to take a formalistic position on technical issues, often referring such issues to be resolved among the parties themselves. They might take similar kinds on decision on future water tariffs issues.) Shirley et al. also cited the limited room for regulatory discretion as another credible commitment gesture by the government. Another critical factor that also enhanced the government's credibility vis-à-vis the water companies was a means tested subsidy, which reduced political opposition to tariff increases. It was estimated that 60% of the bill for the first $20\,m^3$ of consumption by low-income household was paid by the government. From 1990 to 1996, the number of subsidies grew 20-fold from 21,824 to 442,524, costing the country up to US$30 million. Institutional safeguards, such as the difficulty to overturn the set of laws and legislation that underpin the contracts, also enhanced the government's credibility.

EMOS Reform Outcome

Quantitative improvements. Shirley et al. presented evidence that the reform had led to significant observable improvements in the performance of EMOS, despite the already relatively good base that EMOS had begun with:

- EMOS enjoyed an increase in revenue that allowed it also to invest more annually in real terms, so as to extend its water and sewerage connections while maintaining an ageing system. Using 1996 as reference, investments in 1996 (about US$48 million) were more than four times those of 1989, peaking at almost US$60 million in 1992.
- EMOS's market coverage for water reached 100% (from 98%) soon after reform. Its sewerage market coverage improved from 88% in 1990 to 97% from 1994 onwards. These improvements were in spite of the increased number of connections due to more accelerated public housing construction, as well as an expanded area of coverage to include poor municipalities on the fringe of Santiago.
- Increased revenue allowed EMOS to improve financing of its maintenance work, reducing pipe breaks/kilometre from over 0.52 in 1988 to 0.31 in 1994. The increasing trend of unaccounted-for-water was consequently reversed, dropping from almost 40% in 1987 to about 20% in 1996.

- The increased tariffs helped reduce monthly consumption from about 35 m^3/connection in 1988 to around 30 m^3/connection in 1998. Shirley *et al.* attributed the reduction also to widespread EMOS public education about water conservation, as well as the increase in public housing construction that reduced shared housing (where the propensity for an individual household to save water is lower).
- Productivity increased as the number of workers/1,000 connections dropped from 2.1 in 1989 to around 1.76 in 1996. EMOS's total productivity factor maintained almost constant from 1988 to 1996 with inputs at around 60% of outputs, although these did not take into account the value of improved service quality.

Effects on welfare benefits. Shirley *et al.* compared the net benefits from the reform with a counterfactual model assuming no reform had taken place (projecting the factual scenario for 2 years after the end of their observation in 1996, so as to obtain a 10-year 'factual' time period). Their cost–benefit analysis showed the following:

- From 1988 to 1998, the reform generated cumulative domestic gains (include welfare gains by the government, domestic investors, workers, and consumers) of about US$214 million net present value (NPV) in 1988 prices. This was equivalent to about 52% of EMOS's 1988 sales every year in perpetuity. Expressed using 1996 prices, the gains were approximate US$64 per capita (NPV).
- Distribution-wise, the largest gains accrued to the government. At US$185 million in 1988 prices, this was equivalent to 44% of EMOS's 1988 annual sales in perpetuity. Consumers, in comparison, gained only US$3 million (NPV, 1988 prices), which was about US$0.40/connection. The low gain was largely due to the price increases. However, the methodology could not take into account non-quantifiable benefits, such as gains from higher water pressure and reliability. Workers had gained by about US$29 million (NPV, 1988 prices) due to higher wages, which worked out to an NPV of about US$1,710/year in real terms. The gain by the few investors in EMOS came up to about US$0.7 million (NPV, 1988 prices).

Overall, the assessment of Shirley *et al.* about EMOS's reform was positive. While EMOS was eventually privatized in 1999, this privatized period fell outside the scope of their study. For the purpose of key lessons to be drawn from this paper, it also suffices to terminate the assessment at this point.

Chile's Water Code

For the benefit of readers who are new to Chile's water issues, it is useful to know of a larger on-going debate about Chile's Water Code, which governs critical issues of water rights trade. While there are opinions proclaiming the Water Code as a success in a free-market approach to water management, there are also counter-views that question the assumptions underpinning such positive assertions. Bauer (1998, 2004) and Hearne (1998) have carried out extensive empirical and literature research on Chile's Water Code and argued about its pros and cons. Bauer, in particular, had pointed out the lack of water-trading activities in most parts of Chile, as well as the lack of evidence showing whether the Water Code had helped manage issues such as social equity, environmental protection, river basin management, coordination of multiple water uses, and resolution of water conflicts. That debate is another broader study on its own. As pointed out above, the present paper is not intended to contribute to the debate, although the picture would, however, be incomplete if one disregards the fact that, ultimately, the effectiveness of an urban water management system cannot be entirely de-linked from the overarching water rights trade issue, particularly if eventually there is a need to manage better water basin usage among urban and other uses.

The Singapore Experience

Tortajada (2006) gave a concise and comprehensive overview of the Singapore's water management strategy.

Political Context

Tortajada pointed out that Singapore, although well-endowed with heavy annual rainfall of 2,400 millimetres, has not been able to store this rainfall due to land scarcity. Besides relying on natural domestic water catchment, Singapore imports its raw water at about S$0.01/gallon from the neighbouring Johor state of Malaysia, under long-term agreements signed in 1961 and 1962. These were due to expire in 2011 and 2061, respectively. While Singapore was keen to ensure its long-term water supply security well beyond 2061, both sides have not been able to come to agreement on Malaysia's demand for a much higher price for its water, which has varied from 15 to 20 times the current price. Singapore's position was that it has no problem with a higher price *per se*, although its main concern was how the price revision will be decided. Both sides are locked in a stalemate on this issue. Nonetheless, driven by the uncertainty of raw water supply continuity, Singapore was driven to explore pragmatic solutions to increase water supply security and self-sufficiency post-2011. Tortajada identified as a main reason for Singapore's water management success the country's ability concurrently to emphasize 'supply and demand management, wastewater and stormwater management, institutional effectiveness and creation of an enabling environment'. These are elaborated below.

Supply Management

Besides importing water from Johor, Singapore expended much effort to protect its current sources and to expand available sources through desalination and reuse of wastewater and stormwater. Efforts were also made to leverage on technological advancements to increase water availability, improve water quality management, and lower production and management costs. The emphasis on catchment management could be seen from measures to protect the catchment through demarcation and gazetting of such areas, and banning pollution-causing activities in these places. The 1976 Trade Effluent Regulations also strictly regulated the quality of wastewater discharges to streams. Desalination became an important alternative water source, with the opening of the Tuas Desalination plant in 2005 that produced 30 mgd (million gallons per day) to supplement the approximately 360 mgd already consumed. Singapore also invested in technology to recycle wastewater. The Public Utilities Board (PUB), through its 100% sewer connection, was able to collect, treat and reclaim wastewater on an extensive scale. The reclaimed water was termed NEWater. With purity higher than tap water, NEWater is used for industrial manufacturing processes such a semi-conductor wafer fabrication, which require ultra-pure water. A small amount is blended with raw water in the reservoirs, which is then treated for domestic use. PUB projected that Singapore could produce 65 mgd of NEWater by 2011, which would meet 15% of Singapore's projected needs then. In terms of production cost, NEWater is less than that of desalinated water (S\$0.30 versus S\$0.78/m^3). Future water demands would be met with more NEWater, rather than by desalination plants.

Demand Management

Tortajada highlighted how PUB had concurrently emphasized demand management through various well-thought-out policies, such as revising upwards the tariff rate, water conservation tax and water-borne fee, with the aim of reinforcing the water-conservation message. With effect from 1 July 2000, all domestic consumption usage up to 40 m^3/month are charged at a uniform rate of S\$1.17/m^3. Domestic usage above the 40 m^3 level is charged S\$1.40/m^3. The earlier practice of preferential rate of S\$0.56/m^3 for the first 20 m^3 of domestic consumption, S\$0.80/m^3 for 20–40 m^3, and S\$1.17 for anything above 40 m^3, was terminated. Interestingly, non-domestic usage is charged at a flat rate of S\$1.17/m^3 before and after tariff revision. Similarly, from 1 July 2000, domestic consumers would have to pay 30% of tariff for consumption below 40 m^3/month, and 45% for consumption above 40 m^3/month. This compared with zero per cent for the first 20 m^3 and 15% for anything above 20 m^3. Non-domestic usage remains at flat rate of 30% before and after revision. Clearly, the conservation message is targeted at domestic users, with 'penalties for over-consumption'. Similarly, the water-borne fee was also increased from S\$0.10 to S\$0.30/m^3. The effect was a steady decline in domestic consumption from 172 litres/day per capita (lpcd) in 1995 to 160 lpcd in 2005. While the government increased tariff, it also put in place targeted help for those on low incomes, based on housing flat types. Those living in smaller flats receive higher rebates during difficult times, while the hardship cases receive social financial assistance from the Ministry of Community Development, Youth and Sports. Such an approach towards subsidy is more efficient in socio-economic terms compared with subsidizing usage of the first block of consumption, as subsidies end up being leaked to those who do not need it as much as the lower-income group.

Overall Governance

Tortajada complimented the overall governance of water supply and waste-water management in Singapore as being exemplary, in terms of performance, transparency and accountability. The critical areas identified were as follows:

- Human resources: PUB's remuneration and benefits package is pegged to the Singapore Civil Service, which is benchmarked to the market. This ensured that salaries are competitive with the private market, and helps PUB recruit and retain good staff. There was also a system of pro-family human resources policies and staff-development programmes, backed up by a sound appraisal system. This compared favourably with many utilities in Asia, which often had problems with staffing as staff are often recruited based on connections, even when most of them are not qualified for the job. This often led to over-staffing and poor management. Poor pay packages also meant that utilities find it hard to recruit good people in the first place.

- Corruption was a non-issue, as is generally the case in the Singapore Civil Service, which has a strong anti-corruption culture, good remuneration packages and functional institution.
- Autonomy: PUB enjoys much autonomy, with good political and public support, which had made it possible for PUB to increase water tariffs, water conservation tax and water-borne fee to levels that could (1) impact on conservation and consumption behaviour, as well as (2) generate sufficient income to maintain the existing system and invest in future activities. Significantly, tariffs have not been increased since July 2000. In water utilities in developing countries, a lack of autonomy (among other reasons) had led to a steady decline in internal cash generation, from 34% in 1988 to 10% in 1990, and to 8% in 1998. It was also noted that PUB had used the private sector extensively where it did not have expertise, such as desalination and wastewater reclamation.

Overall Performance

In terms of overall performance, PUB invariably appears in the top 5% of all global urban utilities regardless of the performance indicator used. For instance, the entire population has access to drinking water and sanitation; the entire water supply system is fully metered; unaccounted-for-water was 5.14% in 2004; the monthly bill collection efficiency rate was 99% in 2004; and the number of accounts served per PUB employee was 376 in 2004. Tortajada further recommended that Singapore's experience be seriously considered for adoption, after appropriate modifications, by developing countries concerned as well as the donor countries if Millennium Development Goals relating to water were to be achieved. Even more perceptively, Tortajada also identified the following balances that were achieved:

- Water quality and quantity considerations.
- Water supply and demand management.
- Public sector and private sector participation.
- Efficiency and equity considerations.
- Strategic national interest and economic efficiency.
- Strengthening internal capacities and reliance on external sources.

Drawing Parallels

The two examples are clearly different in their political, social and economic backgrounds. But going through the cases, some parallels between the two that may have contributed to successful change could also be discerned. This section will try to draw out these similarities.

Political Impetus

One common key factor identified is the presence of a strong political will in both countries' pursuit of an efficient and sustainable urban water management system. In Chile, the likelihood of reform of its urban water management system was low, as pre-reform conditions, while not ideal, were unlikely to provoke any fundamental changes in EMOS. However, a new Pinochet Administration was intent on privatizing EMOS as part of its overall market-oriented approach towards governance. Events, however, led to the eventual reform of EMOS under public ownership under the Aylwin Administration. The conclusion that could be drawn is that without a strong political impetus, no reform in the national water legislation (and in EMOS) would have taken place in 1988.

In comparison, Singapore's water management system would not have achieved the standard that is seen today had it not been for the political will of its government. A weak government, without the will to attain self-sufficiency in water, could have succumbed to external political pressure. Indeed, if Singapore were to depend on the current supply from Singapore's water catchment area and the 1961 and 1962 Water Agreements with the State of Johor in Malaysia, Singapore would probably still have enough water. However, the political decision was made to increase and diversify Singapore's water resources, so that by 2061, Singapore could be self-sufficient in water. A strong political mandate was what had enabled the country's water authorities to embark on a mission that pulled together resources from different national agencies, and coordinated a national water policy that eventually ensured Singapore's water self-sufficiency. While Tortajada had rightly identified PUB as a key agency that drove Singapore's water self-sufficiency drive, the effort would not have succeeded without cross-agency coordination driven by strong political commitment. According to Tan *et al.* (2009) the setting up of the Water Planning Unit under the Prime Minister's Office in 1971 to study the scope and feasibility of new conventional sources and unconventional sources (water reuse and desalination) led to the Water Master Plan in 1972. This was to become a blueprint that guided long-term development of water resources in Singapore.

Efficient and Effective Institutions

Political will and vision alone would not have been able to deliver a good water management system without the support of an efficient and honest civil service administrative system. Both Chile and Singapore had in place a civil service system known for its efficiency and integrity. In the Chile case study, without a group of able and honest civil servants, the government would not have had sufficient capacity and credibility to carry out its reform. In fact, if Chile had gone the way of privatization right from the outset, it would have been all the more important for the country to have a credible and strong civil service that could put in place a regulatory body (the SSS) effectively to regulate the private sector, and not be captured by the market. Furthermore, given that Chile's Water Code and water legislation strongly involved the roles of the legislature and judicial, it was also to Chile's credit that it had a credible and strong legislative and judiciary arm, that could further bolster the credibility and commitment of the government to its reform process. There was also the practice in Chile of paying officials working in the water utilities adequately, so that the gap with private sector pay was kept small.

Singapore's civil service is generally acknowledged to be an efficient and effective outfit. Just as Chile had paid its EMOS officers well, Singapore believed that paying its civil servants and political office-holders well was a key way to keep corruption level low, as well as to ensure that the system continues to attract and retain quality people.

This principle was applied in PUB. Also, given the cross-sectoral nature of water policy implementation, it was important that agencies were able to work together in a professional way. The Water Master Plan in 1972 translated into integrated planning across various other agencies such as the Urban Redevelopment Authority (URA) and the Housing and Development Board (HDB). Such coordination allowed for long-term planning and proper zoning, which made projects such as the Lower Seleter-Bedok reservoir scheme possible. It would otherwise have been impossible to set aside land for the catchments, stormwater collection ponds system and the reservoir itself. The cooperation also extended from domestic to industrial uses. Tan *et al.* (2009) had highlighted how the Jurong Town Corporation (JTC), the Economic Development Board (EDB) and PUB had worked together to put in place and enforce requirements for industries to seek approval for water consumption above 500 m^3/month.

A People Who Accept Water Pricing

Another key similarity that surfaced was the lack of ground opposition to payment for connection to clean water and sewerage system, allowing the water utilities to be self-sustaining. Pricing water realistically is necessary for a self-sustaining water management system. In Singapore, while the man-in-the-street would in some way be impacted by rising water tariffs and water conservation tax (as evident in reduced consumption), any unhappiness has not translated into political objection, partly because of the targeted subsidy approach adopted by PUB. It could also be because water scarcity is a real threat that the country faced. Besides lacking the land area to catch rainfall, Singapore also lacked aquifers. Singapore was ranked 170 out of 190 in terms of fresh water availability (UNESCO, 2003). While most young Singaporeans have not experienced the water shortage situation in the 1960s (when water rationing had to be practised), all would have been sensitized to the commodity's scarcity.

In Chile, it was significant that EMOS did not face much opposition when it eventually had to go ahead and increase tariffs. This indicated a population that understood that piped water and sewerage system come with a price, and their willingness to pay for it. One possible reason is the strong traditional emphasis on water rights and the tradable nature of such rights have inculcated in the people the notion that water has a price. Such a mentality would have made the task of reform much easier for the reform process.

Ensuring Affordability

Underpinning the success of the water utilities in collecting tariffs is also the key issue of ensuring affordability for those who have difficulties paying for the services. In Chile, the reform was concurrently supported by the launch of a targeted subsidy programme for water and sewerage. The aim was to ensure that no family pays more than 5% of its income for water services. The subsidy is means tested. However, due to the wide income disparity in Chile (Gini coefficients are 54.9 for Chile; 42.5 for Singapore; and 40.8 for the United States; UNDP Human Development Report 2007), some would also leak over to the middle-income group since the gap in classification was not great. This had further reduced opposition to the tariff restructuring (increase) across a broad range of voters and made it more palatable. However, it was important to note the enforced co-payment nature of Chile's subsidy programme. Consumers would have to pay their share of the payment before they could enjoy the subsidy. The co-payment nature of the subsidy ensured that consumers value their water resource and would conserve it.

In Singapore, a targeted subsidy approach was also used to help those who had problems paying their water bills, although the selection criteria in Singapore's case was based on flat/housing size. Known as Utilities-Save (U-Save), these are essentially a quasi-cash subsidy, which is credited into the qualified household's utilities account, to be drawn upon at any time to pay its utilities bills, including water (Tan, 2009). If the rebates are not used up in a month, they remain in the account and can be used in subsequent months. This fixed-quantum pseudo-cash subsidy encourages conservation, since savings from consumption will accrue directly to consumers while they would still have to pay for over-consumption.[3]

Pragmatism and Innovation

Both governments showed a large degree of pragmatism, and did not become stuck with ideological debates or mindset. In Chile, the Aylwin Administration had weighed the political pros and cons of reforming EMOS. While reforming EMOS could potentially erode its political platform, it also saw overall net benefits that a reformed (and presumably profitable) water utility would bring the Administration and the country. One might be sceptical and say that Aylwin's hands were tied by the water legislation to pursue the path of privatization/reform. However, experienced politicians and bureaucrats could easily have exercised innovation by throwing in bureaucratic impediments and dragging their feet in implementing the legislation, if they wanted to. Pragmatism, assisted by innovation, prevailed in this case. Aylwin was able to carry through the reform of EMOS without undermining his own political platform by repackaging it as a reform under public ownership. More importantly, he found congruence with his political platform by combining it with greater social spending, while justifying the project for its fiscal soundness.

Singapore's approach towards managing its water system and supply was similarly a pragmatic one. It showed through its search for alternative sources of water. The notion of consuming recycled water was psychologically off-putting, but the government proceeded with its NEWater project. It managed to get the population's buy-in through intense public education on the potability of NEWater. In times to come, recycled water would eventually become a key 'tap' for Singapore. Likewise, the decision still to keep open the option of having the other 'tap' from Malaysia, despite projections that Singapore would eventually be self-sufficient in water supply through NEWater, is another exercise in pragmatic reasoning. So long as Malaysia was still willing to sell Singapore water at an acceptable price, there was no good reason for Singapore to turn off this 'tap' voluntarily.

Within Singapore's context, innovation in expanding its sources of water supply was key to the success of its water policy. According to Tan (2009), new ways were found to 'catch' water. From protected catchment, Singapore expanded to unprotected catchment, urbanized catchment (which even today is not common around the world). With improved technology (such as membrane technology that has allowed one to treat river water in more built-up areas), the catchment area was increased through the Marina Barrage and the on-going Punggol-Serangoon Reservoirs project. Collectively, these efforts have increased the total water catchment area and will increase to two-thirds of Singapore's total land area of 700 km^2. The search for more 'taps' of water supply was extended to desalination and water recycling. While much progress had been made in recycling, desalination could be constrained by the huge power consumption required to treat the high salinity of seawater. This could be the next challenge for PUB since fresh water currently makes up about 3% of the available water source globally. The world may eventually have to depend on seawater for consumption. However, Singapore had already succeeded in developing a small variable salinity plant that can treat both brackish water and seawater, and harness water in the fringe catchment areas around the perimeter of the island.

Chile, in contrast, is well endowed with water. EMOS would not face the kind of acute shortage that Singapore faces, except when perhaps Santiago's water demand exceeds what its water rights could supply. However, this is a domestic issue and Chileans would have to innovate within the context of their water rights legislation and negotiations among the water rights owners to resolve this shortage problem.

Most Critical Factor?

It is hard to say which of these factors is the most critical. But when one looks at the two cases, one can not help but note that without a strong political impetus as the key driving force, no change would have taken place. The Chilean government could have trudged along with a less-than-ideal urban water management system and probably still lived with it. The Singapore government could have yielded to the wishes of the State of Johor and Singaporeans would have still obtained their water. However, strong political will led the change in both cases. And indeed, one may ask how many countries have not even seriously got off the mark in pursuing a long-term sustainable solution for their water problems. One would have to qualify that without the other supporting factors, the reform or coordination process would have been much more difficult, if not impossible. But these other factors would have been insufficient to provoke or lead change.

Conclusion

It was seen how Chile had built up a more efficient urban water management system through the reform of EMOS, running it as a publicly owned 'private' entity. The circumstances that Chile and Singapore faced were very different when they both embarked on securing a more efficient and sustainable water management system. However, one could distil some common traits from the two urban water management experiences, such as a strong political impetus for change/improvement; strong institutional and bureaucratic support; a situation where water pricing was accepted, supported by an approach towards subsidy provision that encouraged water conservation; and a strong sense of pragmatism and innovation in the way the people approached the water management issue.

Political impetus is arguably the key driving force for change, with other factors playing key supporting roles. Different issues could give rise to the political driving force. In both cases, political/national crises provided this spark that led to a positive change. This paper by no means suggests that one create a crisis in order to advance any water management reform agenda. But one should be alert to the conditions that could generate the political push to bring the issue forward.

There have been many hard analyses of pricing/subsidy policy, water technology, distribution systems, etc. that give ideas for a successful and sustainable water management system. There have also been many convincing analyses and projections of the problems facing urban and rural water systems that should sensitize policy-makers to water as an important issue. But what will shift water to the top of their priorities and sustain their interest in this important issue?

Incentive-wise, a good water distribution and sewerage network contributes to the overall welfare of a country's people. It improves the public health standard, leading to higher product of labour. All these should be sufficient incentive and justification for a country to put high priority on the issue of water. But why has successful reform not seriously taken off in so many places? Hopefully, it does not have to take a crisis in each case.

Notes

1. 'Potential clients' excluded the households in informal settlements without connection to the water/sewerage networks, and the poor communities located just beyond its concession boundary.
2. Chile ranked 23rd out of 180 countries in a 2008 Corruption Perception Index table compiled by Transparency International (2008).
3. Some leakages of such a subsidy to some of the better-off is inevitable. This is a result of Singapore's housing policy, whereby one can still find well-to-do households staying in government flats. Some of these were originally lower-income group when they purchased the subsidised housing, but subsequently progressed to levels of higher income.

References

Bauer, C. J. (1998) *Against the Current: Privatization, Water Markets, and the State in Chile* (Dordrecht: Kluwer).

Bauer, C. J. (2004) *Siren Song—Chilean Water Law as a Model for International Reform* (Washington, DC: Resources for the Future).

Hearne, R. R. (1998) Institutional and organisational arrangements for water markets in Chile, in: K. W. Easter, M. W. Rosegrant & A. Dinar (Eds) *Markets for Water: Potential and Performance*, pp. 141–157 (Berlin: Springer).

Shirley, M. M., Xu, L. C. & Zuluaga, A. M. (2000) *Reforming the Urban Water System in Santiago, Chile*. Policy Research Working Paper (New York, NY: The World Bank).

Tan, Y. S., Lee, T. J. & Tan, K. (2009) *Clean, Green and Blue* (Singapore: ISEAS).

Tortajada, C. (2006) Water management in Singapore, *International Journal of Water Resources Development*, 22, pp. 227–240.

Transparency International (2008) CPI table, *Corruption Perceptions Index*, available at: www.transparency.org/policy_research/surveys_indices/cpi_2008_table

UNDP (2007) *UNDP Human Development Report 2007/2008* (Pittsburgh: Palgrave Macmillan).

UNESCO (2003) *The United Nations World Water Development Report* (Paris: UNESCO and Oxford: Berghahn).

Eliminating 'Yuck': A Simple Exposition of Media and Social Change in Water Reuse Policies

LEONG CHING

Lee Kuan Yew School of Public Policy, National University of Singapore, Singapore

ABSTRACT   *Water reuse is an efficient way of managing water resources in cities, but reuse policies have often been derailed by the 'yuck' factor. While yuck has often been thought of as a problem of public acceptance, this paper argues that there is a more profitable way to frame the issue—as a form of social construct by the media, forming the basis of new learning by the public. This is then illustrated by way of an analysis of newspaper articles in Australia and Singapore. Some preliminary implications for norm formation and possibilities for theory-building are discussed.*

Introduction

There is some agreement among water experts that the current lack of water is less a physical lack than the result of poor management or governance. This, in turn, is a function of other factors such as pricing, management, and infrastructure. Within this broad area of water governance, however, there remains a relatively unexplored issue—why do people in cities still face a water shortage when they can use *recycled water* for drinking? This paper examines one specific implementation obstacle to water-reuse policies—the commonly cited 'yuck' factor.

Although this human aversion, the visceral yuck, is a well-recorded psychological fact, little has been written about it in connection to the use of water-reuse policies. This paper takes a cross-disciplinary approach. From the field of water governance, it can be seen that this yuck factor has been a fairly intractable problem in the implementation of water-reuse policies. From the field of institutional theory, it can be seen that yuck can be usefully thought of as part of the social norms and customs—the informal institutions—surrounding water issues. From media studies, one takes the useful insight that media both constructs and reflects social norms. Influencing and changing media content, therefore, is a way of changing informal institutions.

This paper pulls together different approaches to give a simple exposition of how norms are formed in water-reuse policies, how the yuck factor is presented by the media, and how these affect the implementation of such policies. Two cases are studied and compared: Singapore, which has successfully implemented a water-reuse policy, and Queensland in Australia, whose initial attempt at such a policy failed and is now making a second one.

An analysis of newspaper articles over a period of 12 years (1997–2008) forms the basis of this research.

Some objections to this approach immediately come to mind. First, Singapore and Australia are so different in terms of their political regimes and media composition that any comparison would not yield any practical policy insights. But this paper does not purport to offer general solutions to implementation problems, or even to the limited problem of tackling yuck across different countries. Rather, it offers a framework that allows one to see how the yuck factor can be understood as part of institutional theory and to tackle this problem seriously and profitably. Also, this exposition is limited in the sense that it does not compare or study the influences on the two media, but merely how these two media construct social norms in water-reuse policies.

Overall, the theoretical aim of this paper is to tie the issue of water-reuse implementation to research on informal institutions and use this framework to examine how such informal norms can change.

Water Institutions and How They Change

In examining the question of institutions in the water sector, Douglass North's framework of institutional change (North, 1996) is instructive—he conceives of institutions as structures that humans impose on their interactions. He says that:

> institutions are the humanly devised constraints that structure human interaction. They are made up of *formal* constraints (rules, laws, constitutions) and *informal* constraints (norms of behaviour, conventions, and self-imposed codes of conduct), and their enforcement characteristics. (p. 344)

Together, he says, these institutions define the incentives in society.

North himself has presented a theory of institutional change in which he says that the 'agent of change is the entrepreneur, the decision-maker in organizations'. He argues further that 'it is usually some mixture of external change and internal learning, that triggers the choices that lead to institutional change'. Within institutions, formal and informal constraints change differently—formal institutions change because of a change in the process by which such institutions are formed (North, 1993, p. 64):

> As a result of legislative changes such as the passage of a new statute, of judicial changes stemming from a court decision that alters the common law, of regulatory rule changes enacted by regulatory agencies, and of constitutional rule changes that alter the rules by which other rules are made. In informal institutions, change occurs gradually, and sometimes quite subconsciously as individuals evolve alternative patterns of behaviour consistent with their newly-perceived evaluation of costs and benefits.

The agent of change is the decision-maker (or leader) in organizations.

While change appears to happen relatively slowly during 'business-as-usual' periods of policy-making, during times of crises change can happen relatively quickly.

Why does this happen during a crisis? Culpepper (2008) accounts for it by noting that in normal times entrenched institutions are not easily displaced. During crises, however, a large number of players or actors upset the 'cognitive bases' for such institutions.

The search then begins for a new equilibrium, what he calls 'institutional experimentation'. Such experimentation is characterized by 'deep uncertainty' which places a premium on persuasive argument to create new knowledge. This process of creation is a constructivist approach (Finnemore, 1996; Risse 2000; Schimmelfennig, 2001).

Culpepper (2008) deconstructs the process of institutional change into three stages: crisis, experimentation and consolidation. In a study of union negotiations, where new ideas and new facts bring about a change in entrenched positions, such crises are seen as 'common knowledge events', which lead to the 'emergence of shared ideas in a highly contested area such as wage bargaining'. Therefore, it is seen here that during crises there is creation of knowledge and an accelerated rate of learning.

Applying this specifically to the yuck phenomenon in water-reuse policies, the following hypothesis was derived:

Hypothesis: Informal water institutions will change rapidly when there is a sense of crisis, resulting in rapid learning and/or when there is a strong water leader with a persuasive message.

Recycled Water, 'Yuck' and the Media

'To some, the idea of drinking recycled sewage is akin to eating cockroach-chip ice-cream—unthinkable, even if shown to be safe' (*Weekend Australian*, 11 August 2007, Review, p. 30). This argument that public acceptance still has to be won in those areas which have faced problems in implementing water reuse has often been made. But formulating the problem this way is unhelpful. As Stenekes *et al.* (2006) point out, is not known what the reasons are for the failure of water recycling by the water industry, or what this means for water management institutions. What is meant by public acceptance? How does one measure this? And is a unanimous acceptance by the public a necessary condition for the success of water policies?

Some countries have overcome this feeling of disgust, so clearly yuck is not an immutable factor. There are a few empirical studies (e.g. Margolis, 1996) showing a disparity in the percentage of people who support water reuse across different studies. Among those who oppose water reuse, there are those who adopt their point of view based on an inadequate (or erroneous) understanding of the science behind water recycling (Russell *et al.*, 2006). Last, the effect of public education is unclear, but there is some evidence that public communications have an effect on acceptance (Dingfelder, 2004).

In short, the jury is still out about the level of public acceptance needed for 'successful implementation', and whether those people do not accept such policies say so because they are genuinely opposed or because they just do not have sufficient facts. Given this, for water managers to persist in this track of testing public perception, and garnering public acceptance for water-reuse policies is unhelpful. A more profitable route is to look at the norms, customs and beliefs of a community regarding recycled water.

In so doing, the media is a useful methodological partner in two ways. First, the media influences how people think about the issue. Second, under the social constructivist theory, the media itself constructs these very norms—that is, whether and how the media has the power to create knowledge and shape social norms for water reuse. Here it is important to note that norms, being social constructs, nonetheless have strong implications for objective reality. Saleth & Dinar's (2004) study of water institutions as 'subjective constructs' shows that even though 'institutions are subjective in terms of their origins and operations, [they are] objective in terms of their manifestations and impacts' (p. 26).

Recalling Culpepper's (2008) analysis that the creation of new knowledge requires both reason and persuasion, the issue can be examined according to the framework presented in Figure 1. The rows show the general slant or persuasion of the media coverage, whether they represent scenarios that are supportive or not supportive of water-reuse policies. (Persuasion is used here to indicate the degree to which the media coverage shows its agreement to water reuse.) Note that there is a third possibility: that the media coverage is neither aligned nor hostile—this is the neutral media which then sits on the line between the two.

The columns show the tone of the reports, whether they are emotional or rational. This does not mean that the reports themselves are emotional or irrational, merely whether they represent arguments that have scientific or rational justifications. It may be argued that this creates a bias against opponents of water-reuse policies, since rational arguments will necessarily argue for an acceptance of reused water. This, of course, need not be the case—if there is scientific evidence, for example, that drinking reuse water will increase the occurrence of allergies, then there is a rational argument to be made against water-reuse policies.

The first two quadrants represent a preference for the status quo or against the direction of change. In the first quadrant, the media is hostile on the basis of emotions—positions are entrenched not by way of reference to any objective (third-party) or scientific proof, but on the basis of strong feelings for the particular position. In the second quadrant, there is a dispassionate consideration, and a reasoned refusal, to accept change.

The bottom two quadrants represent a support of change or of the policy under consideration. Support in quadrant 3 emanates from emotional appeal, seen very often in the coverage of wars, a phenomenon commonly called 'Rally Round the Flag' (Groeling & Baum, 2008). The fourth quadrant also supports the policy on objective, or scientific, third-party grounds.

There is an important caveat. First, there is a necessarily subjective element in deciding which reports are emotional and which are rational. Generally, it is taken that those reports that reflect an opinion based on a feeling or an unexplained reaction (e.g., a plumber

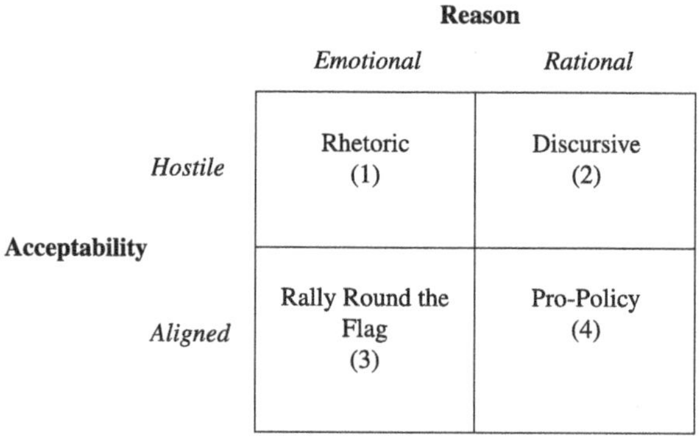

	Reason	
	Emotional	*Rational*
Hostile	Rhetoric (1)	Discursive (2)
Aligned	Rally Round the Flag (3)	Pro-Policy (4)

Acceptability

Figure 1. Media scenarios.

claiming to 'know' the truth about the quality of water) over an expert or scientific opinion) or a partisan or personal attack, rather than policy issues, are emotional rather than rational.

There are then four broad scenarios which are possible for any given policy. Quadrants 2 and 4 are clearly the areas in which most learning would take place, if what is meant by 'learning' the moving towards a new position of consolidation, based on objective facts. Given the above, therefore, the following question is confronted: Which quadrants are most conducive to the learning that will allow an acceptance or overcoming of the yuck factor in water-reuse policies?

A Quick History of Water Reuse in Singapore and Australia

In 1997, Singapore publicly stated that it was aggressively looking at alternative sources of water. There was, in a sense, a crisis engendered by the difficulties with Malaysia over the price of raw water (an issue which remains unresolved today), with the Malaysians threatening to increase prices by at least six times, and with no set formula to peg future increases.

At the time, the country had already been aware of its water shortages for about 30 years—since independence, the country has been importing water from Malaysia of which it had been a constituent part at the point of independence as a colony from Britain.

In 1998, the Singapore Water Reclamation Study (NEWater Study) was initiated as a joint initiative between the Public Utilities Board (PUB) and the Ministry of the Environment and Water Resources (MEWR). The study made explicit that wastewater was being studied as a source of raw water. The water would go through a purification and treatment process using membrane and ultraviolet technologies. It would then be mixed and blended with reservoir water and undergo conventional water treatment to produce drinking water (a procedure known as planned indirect potable use or planned IPU).

By 2001, the PUB released NEWater for non-potable use—wafer fabrication processes, non-potable applications in manufacturing processes as well as air-conditioning cooling towers in commercial buildings. In 2003, the PUB introduced NEWater (about 1% of total daily water consumption) into its water reservoirs. The amount will be increased progressively to about 2.5% of total daily water consumption by 2011.

In Australia, the Queensland State Government initiated the Caloundra/Maroochy Strategic Wastewater Management Strategy in 1997. One of its main tasks was to introduce reused water into the drinking supply. In that same year, the Queensland State Government initiated the Queensland Water Recycling Strategy (QWRS) (Environmental Protection Agency Queensland, 2001).

By that time, water shortages due to drought and long-term, below-average rainfalls have led to a chronic water shortage in Australia. One of the most drought-stricken communities was Toowoomba in south-east Queensland. In 2006, the government held a referendum to recycle waste water to top-up drinking water supplies (Toowoomba City Council, 2006).

Toowoomba would have been the first city in Australia to use recycled sewage for drinking water, with its proposal for a new A$68 million wastewater treatment plant to top-up its water supplies at Cooby Dam. The wastewater would have been treated using reverse osmosis, ultraviolet disinfection and oxidation processes to destroy microorganisms.

But a group of citizens collected some 10,000 signatures for a petition opposing the project (DTI Global Watch Mission Report, 2006, pp. 12–13). In the end, residents of Toowoomba have voted against their wastewater scheme. The Toowoomba poll has put the issue on the national agenda. As the driest inhabited continent, and with droughts increasing in frequency and duration, an intensive dialogue for a sustainable water system is now taking place.

Method

The above section shows that there are good reasons to start the investigation in 1997. Research data on desalination and water recycling in Queensland were obtained from Lexis-Nexis with the search terms 'Queensland', 'recycled water' and 'desalination' from 1997 to 2008. Reports were from five newspapers: *The Australian*, *The Courier Mail*, *The Sydney Morning Herald*, *The Age*, and *The West Australian*.

In Singapore, the database used was 'Newslink', created and managed by Singapore Press Holdings. The search terms used were: 'recycled water', 'NEWater' and 'desalination' from 1997 to 2008. Reports were from *The Straits Times*, *The Business Times*, and *The New Paper* because these are the three major English newspapers in Singapore. Streats and Project Eyeball are now defunct and did not contain any comprehensive reports on Singapore's water policies. The majority of the search results were reports from *The Straits Times*, with a few in *The New Paper* and *The Business Times*.

Each report was tagged according to whether they were supportive of water reuse (positive), hostile to it (negative), or neither (neutral). They were also categorized according to the type of stories—whether news or commentary (all non-news stories, e.g. editorials were considered to be commentaries).

The stories were also examined to see whether and how they portrayed yuck (positive or negative), and how this impacted on the slant of the story as a whole.

Summary of Media Representation of 'Yuck' in Singapore

From 1997 to 2008, there were 223 reports about recycled water in Singapore's newspapers, namely *The Straits Times*, *The New Paper*, and *The Business Times*. Of these, 171 carried positive tones or opinions about recycled water, which the government eventually named 'NEWater'. The positive reports centred on how Singapore need not depend on Malaysia for a long-term water supply, and how safe it was to drink water that was recycled. By the time NEWater began flowing into reservoirs in February 2003, the focus of subsequent media reports was on newer water recycling technologies that could produce more NEWater and at a cheaper price. These technologies later made Singapore attractive to firms looking to do research and development in water recycling.

During the same period, there were nine reports that had negative tones or opinions about recycled water. Many of the reports were related to how NEWater would affect bilateral relations with Malaysia. Politicians in Malaysia took pot-shots at NEWater, warning their people that the water in Singapore could be unclean and even suggesting that Malaysia should sell sewage and not water to Singapore. There was one report in *The New Paper* that said more had to be done to promote NEWater as there was not much awareness about it in the heartlands.

Of the reports, 40 contained the yuck factor, with 29 positive stories. Positive stories included stories about the government assuring Singaporeans that the water was safe to drink, public acceptance, and about how foreigners were giving the thumbs-up to recycled water after tasting it. Negative stories included those about Malaysian politicians and media suggesting that NEWater was not clean and was unsafe to drink.

Analysis of Media Representation of 'Yuck' in Australia

There were no reports about water recycling in Queensland from 1997 to 2000. There were two reports in 2001, when the usage of recycled water in Queensland was first mooted. Recycled water was first used in farms in Queensland in 2001, as it would help save water from the rivers and rain. There were no reports from 2002 to 2004.

From 2005 to 2008, when Queensland became serious about recycling water because of severe droughts, there were 202 reports. Of these, 70 had negative mentions or comments about the Queensland government, compared with 57 positive reports. The rest were neutral. Of the reports, 46 had a negative yuck factor and 36 had a positive yuck factor. Most of the reports did not have any public opinion. The negative reports centred on the safety of treated effluent. Opinion polls showed many Queenslanders felt uncomfortable drinking water that came from sewage and industrial waste. Indeed, some reports said Queensland could not compare itself with Singapore's usage of NEWater, as the latter makes up only 1% of total drinking water, while the Australian state planned to have more than 20% of drinking water comprising treated effluent. Opponents, including the south-west Queensland branch of Commerce Queensland, claimed that no tests were available to detect many of the 100,000 known contaminants of water, including prescription drugs. They also claimed there were no studies of the long-term health effects of drinking recycled water.

Most of the positive reports were about the Queensland government assuring people of the safety of recycled water, or commentaries that recycled water was a solution to the water shortage.

Discussion and Comparative Analysis

On the whole, both media took a fairly even-handed approach to the issue. In both countries, the media were neither campaigning for nor against water reuse. Locating the two cases within the framework shown of Figure 1, Australia belongs in quadrant 1, between 1 and 2, whereas Singapore belongs in quadrant 4. The Australian media held a higher percentage of negative stories (34%) compared with the Singapore media (0.4%). The Australian media also discussed the emotional aspect more frequently. This does not mean that the arguments employed by the Australian newspaper were more emotional or irrational than the Singapore media, only that the number of stories which appeal to emotional elements such as yuck (whether negatively or positively) was higher (a total of 119, making 59%) compared with 34 in Singapore (15%). The amount of political rhetoric was also higher, as many of the stories representing a partisan jockeying of positions.

The Singapore media presented a fairly straightforward picture. The majority of the stories were supportive of water-reuse policies, using rational grounds for justification.

A few descriptive statements can be made about the data below, although a detailed content analysis and corresponding multivariate regression will be able to yield greater insight. First, non-emotional stories tend to have a better chance of being supportive of the policy than emotional ones. For example, those that mention yuck in a negative manner tend to be hostile to the policy.

It is an obvious point but perhaps worth noting that non-emotional stories are not necessarily better for policy-makers (Figure 2 and Table 1)—it depends on the policy. For example, if a policy-maker was advocating a large national celebration, costing millions of dollars, at a time of recession (as Singapore is doing now), it may need an emotional rather than a rational justification. In the case of water reuse, however, it is seen that the rational tone used tend to result in more supportive stories.

Language and Content Analysis

First, and importantly, it is noted that public acceptance played very little part in the discussion. In this sense, this finding squares with the current literature that rhetoric about acceptance is counter-productive in progressing sustainability. It is important to note that the key difference does not lie in the acceptance of one side, and the rejection of the other. On both sides it can be seen that media reports of public attitudes are conspicuously absent.

Hostile versus supportive language. The language employed by the media is very different. In Singapore, the most common adjectives are 'cheap' (or variations such as 'cost less'), 'purified', and 'tried and tested' (for variations such as 'not new', 'track record', 'used in other countries'). In Australia, some adjectives used were 'treated effluent', 'toilet to tap', and 'shit water', the latter used by even politicians who were supportive of the scheme. What reused water is called has implications for the willingness to accept it, as empirical studies have shown. Menegaki *et al.* (2008), for example, have shown that framing treated 'wastewater' as 'recycled water' increased the willingness to use by both farmers and consumers. Language, therefore, is not unimportant to water reuse.

Coherent versus diverse. Singapore's news coverage has consistent key messages. First, water reuse has a good track record and has been used in many other developed countries. Second, political leaders drank reused water openly and frequently. Third, identical words are used by all political leaders. This has the overall effect of reducing uncertainty in the

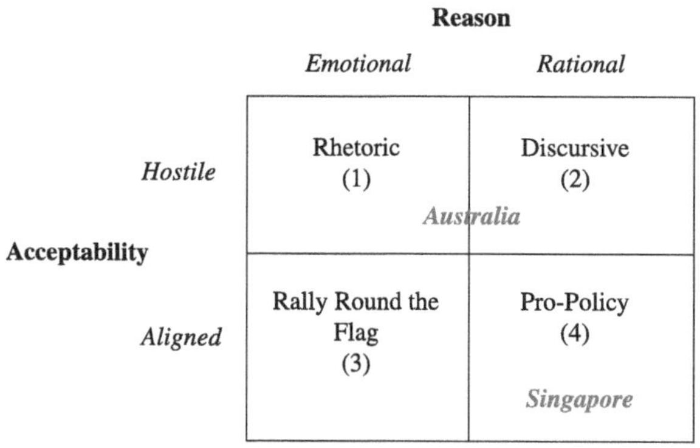

Figure 2. Media scenarios for Singapore and Australia.

Table 1. Summary of reports

	Number of reports	Positive	Negative	Yuck	Positive yuck	Negative yuck	Public opinion
Australia reports	202	57	70	84	35	46	21
Singapore reports	223	171	9	40	29	5	18

public. Australia's coverage, on the other hand, demonstrated a diversity of views on key issues by political players. Emotive language was used often by political leaders. More critically, there was a confusing use of negative terms to describe reused water even by leaders who support the cause. One newspaper pointed out: 'It doesn't help when politicians, both for and against recycling water, confuse the debate by suggesting people will be drinking human waste' (*Sydney Morning Herald* (Australia), 5 September 2005). 'Surely we have to accept that we have to drink our own excrement', was how Chris Harris, a Sydney City councillor and a member of the NSW Greens, phrased it at a recent council meeting convened to discuss Sydney's water shortage (*Sydney Morning Herald*, 5 September 2005). Another example: '"Despite concerns about gender bending (from chemicals in water) there is no evidence of that being associated with what people are drinking", says Ashbolt [Nick Ashbolt, deputy director for the Centre for Water and Waste Technology at New South Wales University]. "It is easy to be alarmist and to raise these concerns."' (*Sydney Morning Herald*, 5 September 2005).

Editorial stance. In both media, the papers themselves did not take an overt stance on the issue. Both media made little mention of public opinion—only about 5% of the stories in the Australian papers and 1% in Singapore reflected this. Given that there were relatively more negative stories (39%) in the Australian media, it can be concluded that this negativity is based on something else other than public opinion. An interesting area for future research is to run a multinomial logistic regression to examine what determines a positive or a negative story. At the moment, it appears that the yuck factor is nested into whether a story is positive or negative. That is to say, there is no negative story with a positive yuck factor and no positive story with a negative yuck. News stories also appear to have a higher probability of being neutral than commentaries.

Change in Informal Institutions

What implications does this analysis hold for the Hypothesis? Recalling that:

> Hypothesis: Informal water institutions will change rapidly when the following obtains either in part or entirely:

- There is a sense of crisis.
- When there is a strong message influencing the beliefs and convictions of people.
- When there is rapid learning.

While the first bullet point was present in both Singapore and Queensland, the second and third bullet points appear to be absent in Australia. That is, while both countries held the necessary condition of a 'crisis mentality' and being ripe for change in its norms, the message was stronger in Singapore, hence leading to rapid learning. In Culpepper's (2008) framework, Singapore appears to have moved quickly to a consolidation phase; Australia is still experimenting and confronting a high sense of uncertainty. Blyth (2001) accounts for actors being stuck in a state of institutional experimentation by saying that they 'lack agreed ideas for making sense of the world' (quoted in Culpepper, 2008, p. 27).

'Never Waste a Crisis'

Crisis can be looked on as an opportunity, when established mindsets can be changed much faster. Thus, in the Australian context, policy-makers will need to review how they can provide what is needed to change this mindset. Locating this discussion within the framework, one would need to move from quadrant 1 to 4 for the kind of knowledge creation that will be most conducive to changing the informal institution of yuck. This, in turn, will need at least the following two processes:

- A coherent communications plan by the water manager (reason), to formulate 'agreed ideas'.
- An engaged media that is neutral at worse, supportive at best (persuasion) to create genuine knowledge.

Further Testing of the Hypothesis

To test these ideas further, the author held interviews with two key decision-makers in the media and the PUB in Singapore for an inside look at the process. The results are produced in some detail below:

A Coherent Communications Plan: Views from the Water Manager

Underlying philosophy. The main goal was essentially to garner public confidence and acceptance. The introduction of a new source of water supply was unprecedented for Singapore, which has all along relied on two traditional sources of water—from local catchments and imported water—since its independence 40 years ago. It was thus critical to ensure that the introduction of NEWater would not meet with any *public opposition*. The main aim was to attain the same level of trust that the population had in PUB water to NEWater.

Public acceptance. Khoo Teng Chye, chief executive of the Public Utlities Board, said:

> The most difficult, yet critical of them all, was to get the public to overcome their psychological barrier towards drinking recycled water and convince them to embrace NEWater as a source of drinking water. To overcome this barrier, a deliberate attempt was made to shift the attention away from the source by focusing on the treatment process, which involves using advanced, state-of-the-art membrane technology. (Interview with author, 24 April 2009).

The PUB tackled the terminology by consciously renaming the terms that had a negative connotation with terms that would better reflect the process or value as a resource.

They did not use internationally recognized terms such as 'wastewater' or 'sewage' because these had a negative connotation.

Stakeholder engagement. The PUB briefed the media and subsequently brought to places in the United States such as Orange County in California and Scottsdale in Arizona to demonstrate to them that water recycling is not a new phenomenon, and that it has actually been a way of life for these people for many years.

The PUB also explained the difference between unplanned indirect potable use which has been practised by cities in Europe for centuries—treated used water is channelled back into the rivers for use by the next city downstream and re-channelled back to the same river for use by yet the next city downstream of it, and this goes on and on—and planned indirect potable use, which Singapore is practising—where the PUB purifies the treated used water to high standards and mixes a percentage of it with raw reservoir water before treating it for the drinking water supply.

The PUB bottled NEWater in attractive packaging so that the public can sample for themselves how pure it is, and these were distributed at grassroots and national events. Top government officials became 'Our NEWater' ambassadors and champions when they were seen drinking NEWater publicly. During the 2002 National Day, some 60,000 Singaporeans toasted NEWater, demonstrating the support and confidence they had in it. The PUB also set up the NEWater Visitor Centre to bolster the public education campaign which was opened by then Prime Minister Goh Chok Tong in 2003.

NEWater terminology. The change in terminology was part of the overall public communications plan to get the public to overcome their psychological fear and accept NEWater. This was also a deliberate effort to minimize the association with terms such as 'sewage' or 'wastewater' which carried a negative connotation.

The PUB official said:

> We also wanted the public to understand that this water is technically not wastewater to be thrown away but water that can be used and reused over and over again, similar to how water recycles itself in nature.
>
> The plants were renamed from sewerage treatment plants to water reclamation plants as they were not merely treating the sewage, but part of the process that reclaims the used water for reuse.

Media as a strategic partner. The current *The Straits Times* editor, Han Fook Kwang, was political editor in 1997. He explained the editorial stance on water issues by stating that:

> water is a life-and-death issue, perhaps more so in Singapore. Strategically, it is also important because it impinges on the broader Singapore–Malaysia relationship. Historically, Singaporeans have had to live through water rationing and the like. Set against this backdrop, it is important for readers to understand how critical it is to for Singapore to have clean water supplies. (Interview with author, 7 April 2009).

He also explained *The Straits Times* interaction with the policy-makers on this topic:

Over the years the PUB, MEWR officials and Cabinet Ministers have all put these issues in perspective on various occasions and in turn we have sought to explain it to our readers.

When NEWater first came about, again, we set about to handle the story like most others, by explaining it simply to our readers. Water reuse is not new and has been used quite extensively and successfully in other countries, including the US. So the 'yuck' factor is not unique to NEWater. By explaining how other countries overcame similar problems when pumping treated sewerage back into the groundwater, rivers or other water supply systems, as well as detailing the science behind Singapore's NEWater, we helped broaden and deepen readers' understanding and acceptance of NEWater. At the same time, we also served as a sounding board to reflect the concerns they had, which the authorities could then address. (Interview with author, 7 April 2009).

The paper also took the important step of ensuring the yuck factor was dealt with in a sensible, rational manner, which was the same attitude adopted by the policy-makers.

Han explains his own perception of the issue:

NEWater was never going to go straight from the plant to the tap. It was used by industry first, and then it was pumped into the reservoirs in a small but slowly growing proportion. All these measures again helped readers understand that NEWater was being monitored and tested closely to make sure the quality of drinking water was always safe. (Interview with author, 7 April 2009).

Instead of downplaying any particular issue, Han felt it important to highlight how different countries deal with their water problems differently and the results of these policies. He added, however, that the paper took care to maintain its own neutrality on the issue, for example, not using PUB jargon and nomenclature. This differentiates *The Straits Times* stance from the 'rally-round-the-flag' coverage, common in stories of the Iraq war by some American media.

Han explained:

As the PUB officially renamed their facilities, we adopted the new names, but we have not stopped using terms like waste water. Depending on the context, we use the most appropriate and accurate terms to help readers understand what are being referred to. (Interview with author, 7 April 2009).

In short, the Singapore media, proxied by *The Straits Times*, appeared to take on a rational, scientific approach to the topic. While recognizing the yuck factor, water as essential good, and the need to overcome the national water crisis, clearly mattered to its editorial stance.

Some preliminary ideas for policy-makers who want to move from quadrant 1 to any other quadrant are the following. In doing so, an easier path, may be from quadrants 1 to 2 and then to quadrant 4, as anti-reuse emotions may be difficult to wear down. As the case of Australia has shown, even the supporters of water reuse get confused into using negative terms themselves when emotions run high. More specifically, moving from quadrant 1 entails that policy-makers themselves move:

- From seeking public acceptance to engaging stakeholders. It is worth noting that neither the Singapore nor the Australian press talked about public acceptance in any comprehensive way. Contrary to common perception, public acceptance in the form of a scientific survey or poll does not appear to be a necessary condition for successful implementation of water-reuse policies. That is, public acceptance as an implicit or tacit agreement appears to work as well. With regards to the media as an informal institution, it can be seen that shaping the nature of this institution is a form of signalling such acceptance. Australia choosing to go the way of a referendum may have set the stage for any future implementation to require such public endorsement. This in turn may stall future efforts.
- From seeking alignment to rational discourse. A strategically aligned media practitioner—the editor of a paper, for example—is quite naturally a useful partner to have. But not having such a partner does not automatically rule out a successful implementation—as long as the dialogue is rational and dispassionate, there is a change to a neutral discourse or 'knowledge creation'.
- From politics to policy. Strong opposition to recycled water was seen as one factor that led the Queensland Premier to abandon plans to supply treated effluent to the south-east region on the eve of state elections in November 2008. A partisan approach, therefore, does not appear to be helpful in creating knowledge in water reuse.

Limits of Analysis

It has been established that the media in both countries constructed the water-reuse issue in different ways, but it is unclear how this difference translated in policy implementation. For example, it is not known if Singaporeans would have accepted NEWater even if the media was very negative about it. At the same time, it remains an unknown if Australians would have opposed recycled water even if the media had been supportive. In other words, what has been possible here is not a chain of causation but merely correlation. More research is needed to determine if such a chain exists.

In addition, this paper's method is a rough scanning of the media reports of the two countries over the past 12 years. While the data are complete (i.e., all the stories on the topic in the time frame have been captured), the analysis is a perfunctory one. To get a better understanding of the editorial stance and the use of rational versus emotive arguments, a detailed discourse analysis of the media content and coding each article should be undertaken. Lastly, there has been no attempt to contact the key decision-makers in Australia's water managers and the media.

Conclusion

This paper has argued that the yuck factor can be usefully thought of as part of the social norms and customs—the *informal institutions* surrounding water issues. It illustrates how the media's construction of water reuse in two countries has had implications for its ability to move beyond the 'experimentation' crisis phase of institutional change. The paper also outlined how the media's role in water-reuse policies can be located within a wider literature of informal institutions, and suggests some preliminary institutional changes that will help the implementation of water-reuse policies.

The contrasting experiences of Singapore and Queensland indicate that, relating to water institutions at least, the process of change in informal institutions can be fairly quick, even in cases when such changes is counter-intuitive. Therefore, some modification in the current thinking on institutional change may be necessary to accommodate this reality. One rich area of research is to explore the work now being done to study the effect of crisis on change. What is the role of crisis in informal institutional change? Is there something intrinsic in water that engenders a 'crisis mentality', thus making informal change faster than it usually is for other issues?

The analysis of the contrasting cases of Singapore and Queensland show how 'yuck' should be tackled in a systematic manner in the pursuit of policy goals. Through the analysis of news content, it can be seen that the media was a key institutional partner in shaping public perception, public learning and, hence, institutional change in water norms.

References

Culpepper, P. D. (2008) The politics of common knowledge: ideas and institutional change in wage bargaining, *International Organization*, 62, pp. 1–33.

Dingfelder, S. F. (2004) From toilet to tap: psychologists lend their expertise to overcoming the public's aversion to reclaimed water, *APA Monitor*, 35 (8).

DTI Global Watch Mission Report (2006) *Water Recycling and Reuse in Singapore and Australia*, pp. 12–13 (London: Department of Trade and Industry).

Environmental Protection Agency Queensland (2001) *Queensland Water Recycling Strategy October 2001* (The State of Queensland, Environmental Protection Agency). Available at http://www.nrw.qld.gov.au/water/regulation/recycling/pdf/qwrstrat.pdf/.

Finnemore, M. (1996) *National International Society* (Ithaca, NY: Cornell University Press).

Groeling, T. & Baum, M. A. (2008) Crossing the water's edge: elite rhetoric, media coverage and the rally-round-the-flag phenomenon, *Journal of Politics*, 70, pp. 1065–1085.

Margolis, H. (1996) *Dealing with Risk: Why the Public and Experts Disagree on Environmental Issues* (Chicago: Chicago University Press).

Menegaki, A. N., Mellon, R. C., Vrentzou, A., Koumakis, G. & Tsagarakis, K. P. (2008) What's in a name?: framing treated wastewater as recycled water increases willingness to use and willingness to pay, *Journal of Economic Psychology*, DOI:10.1016/j.joep.2008.08.007.

North, D. C. (1993) Toward a theory of institutional change, in: Barnett, A. W., Hinich, M. J. & Schofield, N. J. (Eds), *Political Economy: Institutions, Competition and Representation* (Cambridge: Cambridge University Press), pp. 61–70.

North, D. C. (1996) Epilogue: economic performance through time, in: L. J. Alston, T. Eggertsson & D. C. North (Eds) *Empirical Studies in Institutional Change (Political Economy of Institutions and Decisions)* (Cambridge: Cambridge University Press).

Risse, T. (2000) 'Let's argue!': communicative action in world politics. *International Organization*, 54(1), pp. 1–39.

Russel, S., Hampton, G. & Lux, C. (2006) Beyond "information": integrating consultation and education for water recycling initiatives. Paper presented at the Australian Water Association, "From Waters Edge to Red Centre", Alice Springs, Northern Territory, 18–21 April 2006.

Saleth, R. M. & Dinar, A. (2004) *The Institutional Economics of Water: A Cross-Country Analysis of Institutions and Performance* (Cheltenham: Edward Elgar).

Schimmelfennig, F. (2001) The community trap: liberal norms, rhetorical action, and the Eastern enlargement of the European Union, *International Organization*, 55(1), pp. 47–80.

Stenekes, N., Colebateh, H. K., Waite, T. D. & Ashbolt, N. J. (2006) Risk and governance in water recycling: 'public acceptance' revisited, *Science, Technology and Human Values*, 31, pp. 107–134.

Toowoomba City Council (2006) *Water Futures Toowoomba. What Are Our Water Options?* Brochure (Toowoomba, QLD: Toowoomba City Council). Available at: http://www.toowoombawater.com.au/.

Agricultural Groundwater Management in Andhra Pradesh, India: A Focus on Free Electricity Policy and its Reform

RAJENDRA KONDEPATI

Lee Kuan Yew School of Public Policy, National University of Singapore, Singapore

ABSTRACT   *The impact of the free electricity policy on agriculture in the state of Andhra Pradesh (AP), India, is the main focus of this work. It is assumed that this policy has a very high political currency and there is, therefore, a difficultly in recalling it in the short-term. In this context, plausible reforms to this policy are explored with an objective to weed out the inefficiencies in this subsidy regime in the context of groundwater extraction and utilization. These reforms are aimed at reducing the ambit of beneficiaries of this subsidy based on their affordability and increasing the water productivity of agriculture in the state. Some examples exclude large farmers from this policy, offering free electricity conditional upon adopting the System of Rice Intensification (SRI) or adopting micro-irrigation or shifting cropping patterns. These alternate policies are evaluated based on the impact on groundwater extraction, fiscal costs, equity, political feasibility, issues in implementation etc. Finally, it is suggested that the government offers free electricity conditional upon adopting water-efficient cropping practices such as the SRI as a short-term step for increasing the effectiveness of this policy and mitigating its adverse impact on groundwater extraction.*

Introduction

The significance of groundwater in Indian agriculture is reflected by the fact that it accounts for 60% of irrigation requirements (Shah, 2007). Its contribution in the economy has been estimated at 10% of India's gross domestic product (World Bank & Government of India, 1998). Its significance is exemplified by Repetto's (1994) assertion that the Indian Green Revolution in wheat is in fact a tubewell revolution rather than a wheat revolution. Studies suggest that groundwater irrigated farms have 1.2 to three times higher output than those of surface water (Dhawan, 1989). In this context, the promotion of groundwater irrigation has been quite vigorous in the past few decades in India. This in turn has led to rapid extraction of groundwater resources across the country.

The state of Andhra Pradesh is usually referred to as the 'rice bowl of the country' and is a predominantly agricultural state. It also has been facing the deterioration of groundwater, as manifested through its greater emphasis on water conservation efforts throughout the state. The Andhra Pradesh State Water Policy estimates that the increased demand of water will exceed the available supplies of 108 billion cubic meters (BCM) by 2025. One avenue for mitigating the gap between demand and supply is to weed out inefficiencies in the current use of water sources: surface water and groundwater. This paper addresses groundwater consumption in agriculture, as agriculture is its prime consumer.

There are three main sources of inefficiencies in the consumption of groundwater in agriculture:

- Excess agricultural production: in developing countries there are 25–50% losses from the time the agricultural produce leaves the farm gate to the moment it reaches the final consumer (Biswas & Seetharam, 2007), thereby resulting in excess farm production/demand and consequently excess water withdrawal. These losses in turn result in excess production and thereby excess demand of groundwater in agriculture.
- Excess extraction: this groundwater is predominantly extracted through electricity (power)-operated pumps. Any subsidies in the power supply would result in excess extraction of water. A similar flat tariff policy, wherein only an annual charge is levied on the farmer based on the power rating of the electric pump and no further charges for the power consumed, has been implemented in the neighbouring state of Tamil Nadu. Studies on the effect of this policy on the degradation of the water table have been carried out, which indicate that over-exploitation of groundwater resources with this policy is around 30% (Palanisami, 2001).
- Inefficient utilization: another source of inefficiency for groundwater consumption is the usual flood irrigation method adopted by farmers. The traditional irrigation systems can be broadly classified into canal-type surface water based from rivers or groundwater-based furrow type with water flowing from the pumps to the point of irrigation. However, this furrow type uses gravity as a conveying force of water from the source to the utilization point. It is common for a pump to cater to large tracts of land up to 30–40 acres, where different farmers take turns in using the water extracted. Owing to the flow of water over large tracts of land, there is a high seepage loss from source to destination. Moreover, farmers adopt a flood irrigation system, where the water is allowed to flood the entire farm area when it is usually the case that only the plants located in the farm and not the entire farm area require water. Losses are estimated to result in an efficiency rate of only 60% for groundwater-based irrigation systems operating with such groundwater-based systems (Swaminathan, 2006).

As the relevant data for groundwater consumption of Andhra Pradesh in terms of the above inefficiencies are not available, it is assumed that the averages for India also hold good in Andhra Pradesh. This paper addresses the inefficiencies arising only through the excess extraction and inefficient utilization of groundwater in the state, as the other inefficiency caused by losses in the value chain requires a detailed understanding of the infrastructural gaps and policies necessary for addressing these inefficiencies and is beyond the scope of this analysis.

A comprehensive analysis of the policy options addressing water productivity improvements in agriculture is provided by Saleth (2009). It analyses the potential, problems and prospects of water pricing, markets, energy regulations, efficient technologies, water rights and user organizations in promoting water productivity. However, as the overview of this volume presented in its introductory chapter points out, 'the extent and effectiveness of these options are constrained by several institutional, technical and financial factors' (Saleth & Amarasinghe, 2009, p. 21). These constraints are due to the interconnectedness between these options. For instance, the implementation of water markets would require robust definition and enforcement of water rights in place, and so on. This suggests that the above productivity-improving policy options would need some time for comprehensive formulation and implementation. In this context, the emphasis of this analysis is placed on using the electricity tariff as an indirect measure for groundwater conservation in the short-term.

Free Power Scheme and Impact on Groundwater

In 2004, when the ruling Indian National Congress party came to power in the state of Andhra Pradesh, it won the elections on an agenda that it would alleviate the problems of farmers which stemmed from their high indebted situation. As one of the key policy actions in this direction, the government announced this free power (electricity) policy to agriculture, and this policy continued when the same ruling party again won the elections in 2009. This policy, which comes at a high fiscal cost—Rs45 billion, or approximately US$1 billion (US$1 = Rs46.5 as on 6 September 2010) or 4% of the state budget (Government of Andhra Pradesh, 2010)—has been debated widely for its costs and benefits. With the launching of the free power policy, the tariff regime for electricity available to farmers has become a flat rate regime. At this juncture, it would be useful briefly to explore the arguments for and against this flat tariff regime presented in the literature.

Arguments for a Flat Tariff

- Imperativeness: this policy was adopted during a time when there were reports of many suicides amongst farmers who were ridden with indebtedness. As agricultural power costs account for around 15–18% of the gross value of agricultural output (Shah, 2007), waiving the electricity charges offered a timely succour to farmers, resulting in lower costs of cultivation.
- High political demand: given its benefits to farmers, there is great political support for this policy, as exemplified by the return of the same ruling party to power in the subsequent elections and the continuity of this policy.
- Administrative costs: the metered tariff regime has a high administrative cost in terms of recording the meters, billing and ensuring the payment of bills. The administrative costs have been estimated to be around 16% for Maharashtra, which is widely perceived as a progressive state with good governance (Shah *et al.*, 2007). This implies that, in general, the administrative costs of a metered regime could be much higher on average. Moreover, it is commonplace in the state for political parties to promise to waiver agricultural electricity defaults during elections, thus encouraging farmers not to pay their bills when an election is a few months in the future. This process leads to additional transaction costs associated with bill payment defaults and, therefore, a flat tariff policy might be better option in this regard.
- Parity with surface water irrigation: the irrigation canal projects in the state have been built at huge expenditure, thereby subsidizing farmers benefitting from canal irrigation. However, there is no equivalent subsidy for groundwater irrigation, although groundwater irrigation is more productive than surface water irrigation.

This subsidy provides some parity with the indirect subsidy provided to farmers relying on canal-based irrigation.

Arguments for Metered Tariff

- Economically inefficient: as any classical economist would point out, the pricing of water in this regime is far from economically efficient, resulting in no incentives for the efficient allocation of groundwater resources. The power should be priced at its marginal cost of production to reduce distortions in the economy. This will indirectly result in a market-oriented pricing of water, thereby allowing the efficient distribution of water resources. But with a flat tariff regime, water is extracted with no regard to the costs of extraction and therefore this policy results in high economic inefficiencies.
- Regressive subsidies: as shown above, groundwater is over-extracted. This is leading to the deterioration in the groundwater table, requiring high-powered pumps for the extraction of water from deeper sources. The resultant high capital costs could be borne only by rich farmers. Therefore, if this situation prevails, in the medium to long run the policy would only offer subsidies to rich farmers and not the intended needy small and marginal farmers.
- Common property resource: with the offer of a flat tariff regime or free electricity for farmers, both the electricity and the resultant extracted water have become akin to a common property resource. As any common property resource poses problems of over-utilization, groundwater is subjected to over-extraction with a flat tariff regime. A flat tariff policy in agriculture is estimated to result in an additional groundwater extraction rate of 30% (Palanisami, 2001). Excess extraction is also reflected in the increase in the number of newly permitted wells in this state per month, from 87.5 per month before the free power scheme during April 2002–December 2004 to 208 per month during January 2005–January 2006 after the scheme was implemented (Ramachandrulu, 2009). This warrants a closer focus on this policy and necessary efforts to mitigate these negative consequences of the over-extraction of groundwater.
- Fiscal sustainability: any policy should be financially viable for its sustainability in the long run. This policy calls for subsidies of up to 4% of the state's budget and the sustainability of this situation is under question.
- Only monitored can be managed: this refers to the widely used adage that only what can be monitored is what can be managed. Unless one knows the utilization of the resource, it is not possible to manage the resource. Although the resource under consideration is electricity, its generation requires water and therefore impacts water resource management. While the amount of power supplied centrally is known, the lack of regional consumption data makes difficult the region-wise optimization of the power supply and the corresponding distribution of groundwater resources.

Although there were certain restrictions imposed from 2004 in the flat tariff eligibility such as removing the benefits to a few farmers who were also earning non-farm income or mandating that electrical pumps be equipped with energy-saving devices like capacitors, and some seasonal restrictions for crops, these restrictions have been found to have stiff opposition and therefore resulted in inadequate implementation (Birner *et al.*, 2007).

Key Assumptions

As the above arguments point out, the debate of the effectiveness and imperativeness of this policy on electricity supply to agriculture is far from settled. Given the political support to this policy, it would be prudent to assume that this policy will last for a while, until the fiscal situation becomes completely untenable. Therefore, the ensuing analysis does not dwell deeply in this debate. The ensuing analysis considers the existing situation as an opportunity to orient the farm sector in the direction of improving water productivities, leveraging the subsidy offered.

Policy Options

The following policy options have been proposed to address the issue of depleting groundwater resources. These options are chosen with a view to enhance water productivity in the state focusing on targeted subsidies, crop management practices, irrigation systems and cropping patterns:

- Retaining the status quo.
- Excluding large farmers from the ambit of free power policy.
- Free power policy conditional upon water-efficient crop-management practices.
- Free power policy conditional upon water-efficient irrigation systems.
- Free power policy conditional upon shifting cropping patterns.

These options are appraised with respect to the following questions:

- What would be the impact on groundwater consumption efficiencies?
- What would be the impact on fiscal subsidies and their utility?
- Are the subsidies offered progressive in nature?
- How feasible would the policy reform be politically for adoption?
- What would be the prerequisites for the implementation of this policy?

Retaining the Status Quo

In this situation, no modifications are made to the free power policy and the beneficiaries would remain the same. The impacts of the alternate policy options are measured as against this option.

- Impact on groundwater extraction: with an irrigation efficiency of 60% in flood irrigation and an excess withdrawal of 30% owing to the free power incentive, the net extraction levels would be greater than twice the required water levels for irrigation or equivalently more than 50% of net inefficiency in water utilization. In other words, if 12 units of water are required for the current production of a crop, due to an irrigation efficiency of 60%, 20 units are required to be extracted. This is without the incentive of free power. With this incentive there is an additional wastage of 30%. Therefore, the net generated water 26 units or 2.17 times the original requirement of 12 units.
- Fiscal impact: the current subsidy offered by the government stands at Rs45 billion or approximately US$1 billion. But with a net wastage of more than 50% of water extracted, this option results in a huge unwarranted and unproductive loss to the exchequer.

- Equity: the status quo does not place restrictions on the usage of the type of pump while offering free power. However, with a depleting groundwater table, the depth necessary for groundwater extraction increases, hence requiring a higher-power rating for the pump. This increases the capital costs of purchasing higher-capacity pumps. Therefore, retaining the status quo will ultimately skew the benefits offered by this policy towards richer farmers and depriving poor farmers who cannot afford the capital costs of pumps with higher power ratings. Therefore, the equity of this option progressively decreases over time.
- Political feasibility: as evident from the current policy continuity from the previous government to this government, this option has very high political currency, as it is optimal for all sections of farmers. However, given the depleting water table, unless the free power policy offers progressive benefits, there can be political unrest from farmers who cannot afford deeper wells and thus have to use other means of irrigation, incurring higher costs and forgoing this free power subsidy.

Excluding Large Farmers from the Ambit of Free Power Policy

One of the criticisms of the free power policy is that it has regressive subsidies, i.e. the higher the power consumed by the farmer, the higher the subsidy benefit offered. Therefore, this reform addresses this criticism and attempts to make the beneficiaries more specific in terms of their necessity.

Despite a slew of land reforms implemented from the time of independence, the land-holding patterns in the country are very skewed. In the state of Andhra Pradesh, 16.5% of farmers account for a share of 50% of the land holdings (Ramanaiah & Gowri, 2008). In these figures a large farmer is defined as one with average land holdings above 10 hectares (ha). Although free power policy is adopted to offer the necessary support to needy farmers, even large farmers are enjoying this benefit. They get this extra support beyond the economies of scale and bargaining power they enjoy in the marketplace. This policy would reinstate electricity tariffs to these large farmers, with continued free power to other farmers.

- Impact on groundwater extraction: with this policy, 50% of the landholdings owned by large farmers get free power and 50% of those owned by small farmers do not get free power. Therefore, for 50% of landholdings, free power over-withdrawals would be 0%, while the over-withdrawals for the rest of the landholdings would remain the same. Therefore, the net extraction to the requirement ratio drops by the inefficiencies weeded out in the farms owned by landlords owing to free power.
- Fiscal impact: for computing the costs of this policy, wherein 50% of the landholdings owned by the large farmers are not given a power subsidy, fiscal savings would be the savings arising out of tariffs paid by the large farmers and the hitherto lost subsidy to the large farmers. While this policy does not recoup the entire power subsidy given by the state, it recoups the expenditure equivalent to the subsidies given to the undeserving large farmers.
- Equity: in the status quo, the larger the landholding, the larger the subsidy the farmer receives. Therefore, this is a regressive subsidy. In the policy excluding

large farmers from the ambit of free power, equity is restored as only the needy farmers avail themselves of the facility of free power. Although the large farmers also have similar agricultural problems, their cost of capital and other costs are lower due to economies of scale and, therefore, they are usually not in need of such benefits.

- Political feasibility: the smaller farmers in the current regime of free power share certain common problems with large farmers such as the low quality of free power offered to farmers and limited hours of free power supply. In this context, small farmers rely on large farmers to represent their grievances to government officials (Dubash, 2005). Therefore, given the opposition the large farmers would have to this policy reform, it would be very difficult politically for small farmers also to support this initiative owing to their dependence on large farmers. Moreover, there is no apparent redistribution of the benefits foregone by large farmers to small farmers to win over the small farmers. Hence, the political feasibility of this option is very low.

Free Power Conditional upon the SRI System of Cultivation

Periodically, agricultural research has invented water-efficient crop management practices for various crops. However, the adoption of such practices has been lax, as farmers do not know or do not see the incentive of adopting such practices. For example, System of Rice Intensification (SRI) has been known for years. SRI is known to increase the yield of the rice crop meaning an increased agricultural return and a water saving of 30% (Swaminathan, 2006). This policy would make the free power conditional upon adopting this practice or similar water-efficient crop management practices. Adoption of this policy would generate considerable benefits in terms of water conservation, as 86% of the farm area in Andhra Pradesh in based on rice cultivation (Government of Andhra Pradesh, 2008).

- Impact on groundwater extraction: with this policy, 86% of the landholdings that cultivated rice adopt SRI with a water saving of 30%, and 14% continue with free power with the same irrigation systems. This water saving from the rice-growing agricultural belts would reduce the demand for groundwater.
- Fiscal impact: the SRI system decreases water consumption by 30%, resulting in savings to the power consumed by rice farms and the related subsidy. However, this system of cultivation might be unfamiliar to many farmers and therefore the state government needs to provide training using demonstration plots, information sessions etc. The central government as part of the National Food Security Mission offers a subsidy to the states for SRI promotion and therefore the state will not have to bear the entire cost (Government of India, 2007). Moreover, this training offers benefits to society in terms of enhanced yields, food security, water savings and therefore is beneficial in the long run.

Given the conditional instrument adopted here in policy formulation, there would be a monitoring cost to ensure that the farmers availing themselves of the free power benefit stick to SRI. These costs would usually be spread over various governmental bodies. However, to reduce monitoring costs, it is suggested that a mechanism that involves farmers who compete from the region for the same water and power resources be pooled together to form monitoring committees in the likes of user organizations to create a sense of ownership and peer pressure.

- Equity: this policy does not differentiate between large and small farmers. However, during the transition stage from the conventional type of cultivation to the SRI method, the large farmer can always hedge his risk of failure/uncertainty by adopting novel methods only in some of his plots. On the other hand, small farmers take an additional risk, because dividing plots between conventional and SRI method loses economies of scale, increasing costs enormously. Therefore, closer monitoring for small farmers should be considered to decrease their transitional risks. This policy offers free power to rice farmers conditionally and other farmers unconditionally. Hence, to make this policy more equitable, it is imperative that similar methods be researched for other major water-consuming crops in the region like groundnut, sugarcane, coconut and cotton.
- Political feasibility: there would be resistance to this option from farmers in general, given that there is inertia against shifting from the conventional type of cultivation. However, with adequate demonstrations showing the net benefit to them, this option could be feasible. Given that this option demands changes only from the rice farmers, it is possible that they might oppose this initiative on grounds of discrimination. Therefore, there should be active research on exploring how other crops can be effectively grown with higher water-based productivities.
- Issues in implementation: currently the farmers do not adopt this method for multiple reasons: (1) they might be aware of the existence of such method, but owing to their low-risk appetite given the small size of the average landholding in India, they would adopt such practices only after seeing the results for themselves in demonstration plots organized by the government; and (2) although they are convinced that this is a good method to adopt, unless all farmers surrounding a farmer's plot adopt this method, there will be seepage of water from one farmer's plot to his neighbour's at the wrong time in the crop irrigation cycle, which affects the yield negatively.

Therefore, there should be an incentive structure for farmers of a region as a whole to adopt this method. Offering the incentive of free power conditional upon all farmers in a region adopting this method can create this incentive. Amongst farmers, having such common obligations is not a new phenomenon as they are already operating their electric pumps on a common ownership basis.

Free Power Conditional upon Micro Irrigation Systems

The dominant method of irrigation—furrow-type flood irrigation—only results in an efficiency of 60%. However, micro-irrigation systems such as drip and sprinkler irrigation systems, which transport water right to the point of irrigation without any seepage losses, enhance this efficiency to almost 100%. This policy would make the incentive of free power conditional upon farmers adopting such micro-irrigation schemes. The Government of India has formulated a scheme whereby small and marginal farmers are offered an incentive in purchasing such a system with 50% of the cost borne by the farmer, 40% by central government and 10% by the state government (Government of India, 2006). The promotion of this scheme can be enhanced if this policy reform is adopted.

- Impact on groundwater extraction: with this policy, the farmers would adopt drip irrigation systems resulting in a drastic reduction in seepage losses while the over-withdrawal rate owing to free power policy might remain the same. While quantifying the water savings to the state through this reform, one needs to consider that the state of Andhra Pradesh is predominantly a rice-growing area. The benefits from micro-irrigation to rice would not be as significant as offered to other crops, as rice grows densely over a plot of land and therefore flood irrigation is the most optimal method. However, there would be benefits from micro-irrigation even in rice cultivation, as the water has to reach the plot of land travelling through long stretches of land from the pump, which is the source for 30–40 acres of land.

- Fiscal impact: the costs assessment for this policy also has three dimensions: fiscal costs; subsidy costs for owning micro-irrigation infrastructure; and training and monitoring costs as in the above option. In this policy the efficiencies of irrigation system have gone up. To that extent of enhanced supply of water hitherto lost in seepages, water extraction could go down and resultantly reduce power subsidies, creating a lesser burden on the exchequer.

    This policy adoption assumes that the state government is going to promote the scheme offered by central government, thereby willing to bear 10% of the costs of the system. Although the state government need not promote this policy to adopt the suggested policy alternative, the farmer needs a strong incentive to move beyond conventional irrigation systems and the incentive offered by the central government scheme would aid in providing the necessary impetus to the farmer to adopt this system. As in the previous option, there would be additional training and monitoring costs to implementing this option.

- Equity: this policy does differentiate between small and large farmers and offers a subsidy only to small farmers. However, given the novelty of this system, it is only fair that the small farmer be given an incentive to adopt this system. Large farmers can hire expertise to operate these systems or hedge their risks by adopting these systems in some areas of their land.

- Political feasibility: in this option the large farmer is not given any benefit and the small farmer is compelled to shift to micro-irrigation systems. This option can run into political opposition from both segments of farmers. Large farmers might oppose this policy because they do not get any incentive; small farmers might oppose this policy because they might not see any immediate returns and, therefore, might take some time to be convinced that this policy benefits them. This option might face opposition from the agricultural labourers as well, because this system reduces the dependence on labour; however, given the booming national and state economies and employment guarantee schemes, this might not be a pressing concern. Finally this policy involves a change of farmers' modus operandi and at the same time offers them additional benefits, but only with a certain lag, the policy-makers judging the political feasibility of this option should be a bit circumspect.

- Issues in implementation: although it appears that the above scheme floated by central government is a good incentive for farmers to take up micro-irrigation systems, their lack of expertise in operating these systems, which would entail fine-tuning the water pressure and flow rate settings, will hinder them from adopting these systems. Therefore, an element of training is necessary when adopting these systems. With training and experience the ability to handle these systems would no longer be an issue of concern.

Another bottleneck for the adoption of this reform is that the initial capital costs are perceived to be high for farmers and this therefore hinders them from the adoption of the micro-irrigation systems. It is estimated that these costs will be recouped in one to two years (Swaminathan, 2006). However, if farmers still perceive it as a concern, financing could be arranged by mediating with banks on behalf of farmers. Further, crop-specific incentives based on the profitability of the adoption of these systems could work as an added incentive (Narayanamoorthy, 2009).

Free Power Policy Conditional upon Shifting Cropping Patterns

This policy reform would offer the benefit of free power only to those farmers who stick to the cropping pattern prescribed by the government. This cropping pattern would be formulated by taking into account the water productivity of various crops. Although this option might conserve water and resultantly reduce fiscal subsidies, its feasibility and equity depend on the specifics of the cropping patterns prescribed and the profitability of the crops. Malik (2009) highlights an example in the context of a shift in cropping pattern, where the state of Haryana promoted the cultivation of sunflowers in place of high water-consuming rice, and when the adoption of sunflower cultivation crossed beyond a critical level, the profitability of sunflower cultivation came down drastically, compelling farmers to shift back to rice. Therefore, Malik points out that any shift in cropping patterns would require the 'realignment in output pricing policies' to ensure the sustainability of the same. It need not be stated that this realignment is easier said than done and this policy reform might gain traction only in the long run.

Recommendations to Free Power Policy

The above analysis has analysed the impact of four plausible reforms that could be taken up in the existing environment and the analysis is summarized below to come up with a way forward. Offering free power conditional upon shifting cropping patterns seems to be a long-term exercise once the market support systems are sufficiently reoriented. Targeting free power only to small farmers, although they might generate higher revenues, would equally run into heavy political constraints. To make this option feasible, the other common institutional problems, for which the small farmers depend on the large farmers to represent them, such as a lack of quality of power supply, should be resolved first.

Offering free power conditional upon micro-irrigation seems to have very high potential for water savings given that the 40% inefficiencies in the flood irrigation systems would be weeded out. However, adopting this option would mean that the other issues such as (1) a lack of operating expertise amongst farmers and (2) high capital costs and their financing must be resolved first.

Offering free power conditional upon SRI cultivation offers significant water savings to the state, as the predominant area (84%) under cultivation is growing rice. Moreover, this option also offers benefits to farmers in one crop cycle of rice (4–6 months). Therefore, this option is more likely to succeed politically as well financially for the state as well as the farmers. Moreover, this policy increases the yield to farmers offering additional food security. Given these above reasons, this option could be a good reform to the existing policy. In the same light, innovative practices in other crops should also be actively looked at. Even promoting research and development with an objective to enhance the water productivity of crops would yield rich dividends in the long run.

Conclusion

This free power policy is formulated with primary consideration of the farmers' distress in agriculture without giving adequate attention to its impact on ground water economy. It is a product of the often-quoted delineated approach of the current policy-making and calls for a relook with a coordinated approach to the management of the water, food, energy and agricultural sectors. At this juncture, it would be prudent to ask a question. If farmers need water, are there alternate means to provide irrigation without squeezing the energy sectors and creating distorted incentives for indiscriminate ground water extraction? Or what is the optimal way to subsidize irrigation through the extraction of groundwater in an environment where the rights over this precious resource are ill-defined.

The free power policy apparently is not efficient, as it fundamentally lacks any incentive for efficient allocation of energy and water resources. In view of this, the academic literature predominantly has to chose to focus on alternatives such as pricing, rights, user organizations, technologies etc. that have no connection with the existing policy. Although this approach is useful for long-term considerations, it is necessary for analysing feasible options that would bridge the existing policy to the above alternate efficient policies in the short to medium run. Therefore, this paper suggests that until the optimal long-term options are politically feasible, one should view the current context as an opportunity to compel the farming community to orient them in the direction of sustainable agriculture leveraging the incentive of free power.

In this direction, reforms to the free power policy are suggested with a focus on groundwater conservation and these reform alternatives are assessed with respect to their impact on groundwater extraction, fiscal situation, equity and political support. Based on this analysis, it has been suggested that offering free power conditional upon adopting water-efficient agricultural practices like System of Rice Intensification (SRI) should be the way forward when mitigating the impact of the current policy on groundwater extraction. Besides increasing the incomes of farmers, it decreases the subsidy burden on the government. This decreased subsidy burden would release resources for other developmental activities, yet providing the equivalent agricultural support as before. Therefore, it is in the political interests of the government in power to reap the benefits of this policy.

In the long run, it is imperative that the government ensures that the energy provision to agriculture has no quality and service-related issues. This would reduce the coping costs borne by farmers and perhaps enable them to consider favourably changes to the existing free power policy. It is also a reality that policy actions based on water productivity and availability would be necessary for sustainable development across the regions. This would entail discriminating policies like specific incentivized cropping patterns across regions. Therefore, the imperativeness of such customized policies should be promoted in public discourse so as to prepare the public for future policy changes.

In a nutshell, as this paper suggests, an existing regressive subsidy is not necessarily a lost opportunity. The existing subsidy could be used as leverage to formulate policies to direct the resource consumption in the correct direction. Therefore, this approach could also be adopted in other states that have similar subsidy structures. Finally, the adoption of the above conditional policy reforms should be followed by institutional reforms like water rights, markets, pricing, user organizations etc. to ensure sustainable development and the management of water resources as a whole.

References

Birner, R., Gupta, S., Sharma, N. & Palanisamy, N. (2007) *The Political Economy of Agricultural Policy Reform in India* (New Delhi: International Food Policy Research Institute (IFPRI)).

Biswas, A. K. & Seetharam, K. (2007) Achieving water security for Asia—Asian water development outlook 2007, *International Journal of Water Resources Development*, 24(1), pp. 145–176.

Dhawan, B. (1989) *Studies in Irrigation and Water Management* (New Delhi: Commonwealth Publ.).

Dubash, N. K. (2005) The electricity-groundwater conundrum: the case for a political solution to a policital problem, Paper presented at the 4th IWMI-Tata Annual Partners' Meet, National Institute of Public Finance and Policy, New Delhi, India.

Government of Andhra Pradesh (2008) *Andhra Pradesh Statistics* (Hyderabad: Directorate of Economics and Statistics). Available from: http://www.apdes.ap.gov.in/AP%20admin%20Setup.htm (accessed on 27 March 2010).

Government of Andhra Pradesh (2010) *Annual Budget* (Hyderabad: Ministry of Finance).

Government of India (2006) *Micro-Irrigation* (New Delhi: Ministry of Agriculture).

Government of India (2007) *National Food Security Mission* (New Delhi: Ministry of Agriculture).

Malik, R. (2009) Energy regulations as a demand management option: potentials, problems and prospects, in: R. M. Saleth (Ed.) *Promoting Irrigation Demand Management in India: Potentials, Problems and Prospects*, pp. 71–92 (Colombo: IWMI).

Narayanamoorthy, A. (2009) Water saving technologies as a demand management option: potentials, problems and prospects, in: R. M. Saleth (Ed.) *Promoting Irrigation Demand Management in India: Potentials, Problems and Prospects*, pp. 93–126 (Colombo: IWMI).

Palanisami, K. (2001) Techno-economic feasibility of groundwater exploitation in Tamilnadu, in: *ICAR-IWMI Policy Dialogue on Groundwater Management* (Karnal, Haryana: CSSRI).

Ramachandrulu, V. (2009) *Groundwater in Andhra Pradesh in India: The Case of Privatization of a Common* (Secunderabad: Centre for World Solaridarity).

Ramanaiah, M. & Gowri, C. (2008) Land reforms and agrarian conditions in Andhra Pradesh during post Independence era, *Social Science Electronic Publishing*, 20 November.

Repetto, R. (1994) *The 'Second India' Revisited: Population, Poverty and Environmental Stress over Two Decades* (Washington, DC: World Resources Institute).

Saleth, R. M. (2009) *Promoting Irrigation Demand Management in India: Potentials, Problems and Prospects*, Strategic Analyses of the National River Linking Project (NRLP): Series No. 3 (Colombo: IWMI).

Saleth, R. M. & Amarasinghe, U. A. (2009) Promoting irrigation demand management in India: policy options and institutional requirements, in: R. M. Saleth (Ed.) *Promoting Irrigation Demand Management in India: Potentials, Problems and Prospects*, pp. 1–24 (Colombo: IWMI).

Shah, T. (2007) The groundwater economy of South Asia: an assessment of size, significance and socio-ecological impacts, in: M. Giordano & V. Villholth (Eds) *The Agricultural Groundwater Revolution: Opportunities and Threats to Development*, pp. 7–36 (Colombo: CAB International).

Shah, T., Scott, C., Kishore, A. & Sharma, A. (2007) Energy-irrigation nexus in South Asia: improving groundwater conservation and power sector viability, in: M. Giordano & K. Villholth (Eds) *The Agricultural Groundwater Revolution: Opportunities and Threats to Development*, pp. 211–242 (Colombo: CAB International).

Swaminathan, M. S. (2006) *More Crop and Income per Drop of Water* (New Delhi: Government of India, Ministry of Water Resources).

World Bank and Government of India (1998) *India—Water Resources Management Sector Review: Groundwater Regulation and Management Report* (Washington, DC: World Bank, and New Delhi: Government of India).

Index

Page numbers in *Italics* represent tables.
Page numbers in **Bold** represent figures.

Water International

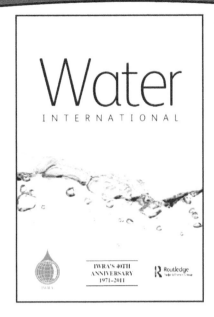

Water International is the official journal of the International Water Resources Association (IWRA), founded in 1972 to serve as an international gateway to the people, ideas and networks that are critical to the sustainable management of water resources around the world. Water International's articles, state-of-the-art reviews and technical notes are policy-relevant and aimed at communicating in-depth knowledge to a multidisciplinary and international community. Water International publishes both individual contributions and sets of papers on cutting edge issues.

To find out how to submit a paper to Water International, visit the journal homepage at:
www.tandfonline.com/rwin
and click on the 'Authors and Submissions' link

Editors:

James Nickum, Tokyo, Japan

Yoram Eckstein, Kent State University, USA

Cecilia Tortajada, Third World Center for Water Management, Mexico

Anthony Turton, Centre for Environmental Management, University of Free State, Bloemfontein, South Africa

Philippus Wester, Wageningen University, The Netherlands

ISSN:
0250-8060 (Print), 1941-1707 (Online)

Routledge
Taylor & Francis Group

International Journal of Water Resources Development

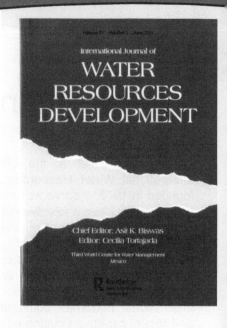

International Journal of Water Resources Development covers all aspects of water development and management in both industrialized and Third World countries. Contents focus on the practical implementation of policies for water resources development, monitoring and evaluation of technical projects, and, to a lesser extent, water resources research. Articles are rigorous and in-depth, and range in approach from applied geographical analysis to the examination of strategic, economic and social issues.

The journal would be of interest to:

- Academics and researchers in the water resources field;
- Policymakers and managers in all organizations that are affected by, or concerned with, water resources development;
- Hydrologist, economist, sociologist, geographers, geologists, meteorologists and limnologist;
- Lawyers and administrators;
- Civil, mechanical and electrical engineers with an interest in this area.

Editor-in-Chief:

Professor Asit K. Biswas Biography, *Third World Centre for Water Management, Atizapan, Mexico*

Editor:

Cecilia Tortajada, *Third World Centre for Water Management, Atizapan, Mexico*

ISSN:
0790-0627 (Print), 1360-0648 (Online)

To find out how to submit a paper to *International Journal of Water Resources Development*, visit the journal homepage at:
www.tandfonline.com/cijw
and click on the 'Authors and Submissions' link.